Animals Impacting the World

Dinah Roseberry

Mary Gasparo

The greatness of a nation and its moral progress can be judged by the way its animals are treated.

~Gandhi

Paranormal Books by Dinah Roseberry
Ghosts of Valley Forge and Phoenixville, **978-0-7643-2633-2, $14.95**
Cape May Haunts, **978-0-7643-2821-3, $14.95**
Spooky York, Pennsylvania, **978-0-7643-3021-6, $14.99**
Spooky Creepy Baltimore County, **978-0-7643-3254-8, $14.99**
Psychic Pets: Solving Paranormal Mysteries, **978-0-7643-3398-9, $12.99**
Ghost Hunters' Tool Kit, **978-0-7643-3912-7, $34.99**

Schiffer Books are available at special discounts for bulk purchases for sales promotions or premiums. Special editions, including personalized covers, corporate imprints, and excerpts can be created in large quantities for special needs. For more information contact the publisher:

4880 Lower Valley Road • Atglen, PA 19310

Published by Schiffer Publishing Ltd.
4880 Lower Valley Road
Atglen, PA 19310
Phone: (610) 593-1777; Fax: (610) 593-2002
E-mail: Info@schifferbooks.com

For the largest selection of fine reference books on this and related subjects, please visit our website at **www.schifferbooks.com**
We are always looking for people to write books on new and related subjects. If you have an idea for a book, please contact us at proposals@schifferbooks.com

This book may be purchased from the publisher.
Include $5.00 for shipping.
Please try your bookstore first.
You may write for a free catalog.

In Europe, Schiffer books are distributed by Bushwood Books
6 Marksbury Ave.
Kew Gardens
Surrey TW9 4JF England
Phone: 44 (0) 20 8392 8585; Fax: 44 (0) 20 8392 9876
E-mail: info@bushwoodbooks.co.uk
Website: www.bushwoodbooks.co.uk

Library of Congress Control Number: 2012950308

Designed by RoS
Type set in Tiranti Solid LET/Book Antiqua
ISBN: 978-0-7643-4237-0
Printed in China

All text and photos by authors unless otherwise noted.
Image Unhealthy swan © Gotcha, yorkie poo cuteness © louge, abstract blue green background texture © Bella D, picture of african animals in savannah at dawn time © NexusPlexus,
Spring Green Leaves border © Subbotina Anna. Unhealthy swan © catcha. Yorki Poo Cuteness © louage.
Shire horse grazing © roger ashford. Photos courtesy of www.bigstockphoto.com
Some photos are courtesy of Art Explosion
Cover images by Chuck Rineer, Jim O'Rear, Renee Harper, Ro
Ouija is a registered trademark of Parker Brothers

UNDER NO CIRCUMSTANCES SHOULD ANYONE TAKE THE MESSAGES HERE AS A REPLACEMENT FOR ANIMAL VETERINARIAN CARE OR MAJOR (OR EVEN MINOR) LIFE DECISIONS.

Dinah's Dedication

To the people and animals that, when tied together by the soul, impact the world.

Mary's Dedication

This book is for all who have taken animals into their hearts.
For those who took the ones others didn't want and gave them a forever loving home.
And to inspire all future animal communicators to go forward...
You are needed.

Animals

Contents

Foreword

Anne McCurdy
Search Team Manager and Canine
Training Director
Indiana Task Force 1
Canine Urban Response Team

I watched the German Shepherd head down another enormous ravine and reminded myself "that which doesn't kill you makes you stronger." I slid down the side of the slippery hill on my butt (for the umpteenth time that day) following a handler and her dog on a search dog evaluation.

The dogs have a limited time period to clear a very wooded and rough search area to look for a missing "victim." I had arrived very early in the morning to hide the poor unsuspecting soul who offered to come out to "help the search dogs." Little did she know she would be dropped off in the middle of the woods with her bottle of water and the instructions, "Lie there with your head down and don't move at all until a dog barks at you."

The handler has an important job in this evaluation. Just getting to the point of testing has required years of training and lots of work. Once they make it this far in the very arduous process, they must be prepared and be able to work the dog, to give it the best advantage of locating a lost person. They must be methodical, but have a sense of urgency. They must have a good sense of direction and know how to use their equipment (GPS, compass, etc.). They must carry what they need - nothing more, nothing less. If they get tired, they must soldier on.

Most importantly, they have to develop a bond with their dog that allows the utmost of trust of each other. The handler must be willing to take calculated risks to allow the dog to search independently, and trust the dog to make the right decisions during the search. The dog must trust the human partner to give him the independence to do his job, but to keep him safe from inherent hazards like getting shot by a hunter or hit by a car.

I had specifically chosen this site for the difficulty of terrain. This was the 4th team I had evaluated that day, so I was getting tired. Each team had searched the area only once, but I had to follow every dog through the entire area. I don't expect every dog to pass this test - I want only the BEST dogs to pass this test.

By this point, the dog I had been following had scaled what appeared to be a cliff the size of Mt. Everest. The handler and I crawled up the side of the mountain attempting to keep up with the dog. Then.... BARKING. Big loud barking. We made it to the top in time to see the dog standing over the victim alerting us that he had found her.

As I evaluate the teams, I ask myself, "If I am lost, would I want this dog to come look for me?" Sometimes my eyes even tear up with pride when I think, "Yep – this one's a keeper."

As Search Team Manager for Indiana Task Force 1, it is my job to be sure the dogs on our team are qualified to do the job. The wilderness test described above is really a "side job" for most of the dogs. The mission of the task force is to provide search and rescue operations in the aftermath of a disaster. Our dogs are the primary "tool" the team uses to search the rubble for missing victims. These privately owned dogs are trained by their handlers to locate live human scent under the debris of a collapsed structure. The dogs will locate the human scent without ever seeing the trapped victim and bark to alert the handler that they have located a live person. Though our team has responded to many disasters, including September 11^{th,} and countless hurricanes, like Katrina in 2005, the need for search dogs extends far beyond the role of "disaster dogs." Children get lost or run away, and elderly patients with Alzheimer's get disoriented and lost more often than you know. Properly trained dogs are the best resource for locating these missing people.

Dogs that also do wilderness search have more opportunities to locate a person alive. Thankfully, disasters don't happen as regularly as people just getting lost. And the probability of finding a person alive is greater if they are sitting in the woods

as opposed to being crushed under the rubble. Our team trains the dogs for many specialties including disaster, wilderness, and human remains detection.

I am fortunate (or possibly unfortunate) to have had the opportunity to watch dogs work in the most adverse of all conditions – the World Trade Center. These dogs worked tirelessly for days with the hope of finding someone alive. Disappointment was evident on every face – human and dog – as they left "the pile" at the end of their shift each day. Day after day of searching and never finding anyone alive takes its toll on everyone, humans and their canine partners alike.

Our dogs are an extension of ourselves. We can communicate with them without even saying a word. We can teach them to perform tasks that we ourselves are not able to do – like sniffing for smuggled fruit or drugs, for skin cancer, or for a person buried deep beneath the rubble. I believe the bond between person and pet can be as strong as any relationship between people.

These dogs impact the world. I hope that you find comfort in the stories and lessons that follow, for animals impacting the world include all sizes of impact. I invite you to join Dinah and Mary on their quest for the truth behind all matter of animal communication – love.

~Anne McCurdy

An Introduction
(of Sorts)

"An animal's eyes have the power to speak a great language."

~**Martin Buber**
German Philosopher

How to begin? This second animal book, written to benefit the people of the world who love the creatures of the world, did not start out as such. It was to be a follow-up to the first book, *Psychics Pets: Solving Paranormal Mysteries.* The idea was to have more experiences in the animal communications arena so that people could both continue to learn the skill and pass it on to other like-minded folks. Readers were to be exposed to heartwarming stories of bonding between animal and human.

Of course, then Spirit takes over and the book written is presented more on a needs basis than a follow-up basis. It didn't take long for Mary and I to realize that though, yes, animal communications and connecting to the wild kingdom were mandatory, this would also be about other kinds of communication - those that impacted the world and those of us living in it. For, you see, not everyone (unfortunately) has a desire for communicating on that higher level with their animals. They feel that:

a. It can't be done,
b. It's too touchy-feely, and
c. Who has that kind of time?

Too, Mary and I were being pulled in varied directions (that seemed odd at first). I was falling ever more deeply into the interactions with the Light Beings [spiritual beings that had me channeling] and UFO/Star Being contacts. Mary was becoming quite the empath, with continual attraction towards animal medical conditions and many ghostly experiences involving the investigations into paranormal phenomena. So of course, when researching a book, things affecting these topics are often brought to the forefront.

For my side of it, out of the clear blue, I'd begun channeling Light Beings. Yes, I know that sounds crazy to some. I'm finding, though, that it's no different than finding out that you have a skill that you didn't know you had and then honing it. My brand of Light Beings are spiritual beings who have been placed with me to teach me how to raise the vibrations of my fellow humans and animals. I could go into great detail about how this works, why it is, and what every human being can do to be prepared for the strange, wonderful, and scary changes that are coming down the road. Suffice it to say that I'm advised on ways to help others rise up against the negativities that would bring their lives "down" by Beings outside my nature.

Mary has been finding that she cannot walk into a restaurant, shopping center, or even a public bathroom without being pelted with the medical or psychological issues of those around her. It's been tough on her. She feels everything. These empathic feelings have been especially present as she struggles to help the cat population that comes in a constant stream through her cat rescue. People, who know how she learned this great animal communication skill and then trumped it with empathic "knowings," are routine "customers," too, whenever an issue occurs with their loved pets. She, of course, always rises to the occasion. So we're incredibly busy in ways that neither of us would have expected just a year ago.

We meld when we use her empathic skills and my channeling of the Light Beings to determine what is happening and how we can help the animals that float into our fields of energy. It has been a strange liaison. Our new directions have changed our initial endeavor to teach all people how to communicate with their animals. Instead, we find that there are stories to be told of learning, dedication, and love between animals and humans that impact the world at large. These stories - whether related to formal animal communications or just simple and regular interactions - make such a difference and provide such a backdrop for love and forward movement for the soul that we had to change our focus to bring you the stories presented here. All have an individual flair that profoundly affects a small group…or a large group…or one individual, but when read by others, these stories send a message so strong to the populace that the effects are truly astounding. We ask you to think about these stories, compare them to your own, and tell them to others in the hope that bonds are further cemented between the animal and human worlds.

We hope that you will enjoy reading these stories as much as we have collecting them. Feel free to send us your impacting animal stories so that we can continue to share the voices of those who would be our best friends!

A Book for the Birds

We'd like to start this book by talking about the seemingly main character of our *last* book. The lovely, if crotchety, Bentley the cockatiel peppered nearly every story, as he was both my and Mary's first strong-willed contact to the other side via an animal on this side. This bird, belonging to the famed Hollywood ghost and witch author Marla Brooks, gave much advice, kept us on our toes, and berated us at every turn while we were learning the animal communications craft. He was a mainstay in what I called my "Core Group" of animals that shared a daily session to discuss events, questions, or anything at all related to any animal who needed us. There were dogs and cats and even a rock that rallied around this small bird with a big voice each day after I'd learned to communicate with animals. Along with Mary's cat, BJ, and my dog, Rufus, our dear Bentley passed away in 2010 after living over 25 years. Great losses all, for those of us in the midst of touching an unknown world.

Photo Credit: Marla Brooks

Not a Deserter

Bentley, however, did not desert me. Though I found that I was not able to actually interact with the animals in the same manner as I'd done prior to the arrival of the Light Beings, Bentley and my protector horse from that time, Solo, continued to be with me. Bentley, however, was no longer as vocal. He'd come to me in a flurry of writing – a caricature figure almost, scribbling on a note pad and looking up with growling eyes as he made his notes. He told me he was writing his memoirs and that all of us should be very frightened about it. I remember laughing and saying to myself: *Good Old Bentley.*

It was difficult for me, though, not to be as close to the animals as before. Other communicators told me that it was a choice I'd made internally. That could be, but I'm not so sure. The Light Beings overpowered and overshadowed all my endeavors, this one included. They had important work for me to do – that of course included both animals and people.

The Light Beings Bring News of Bentley

Marla Brooks, however, was grieving more than any of us, as she'd had her companion for twenty-five years and now he had passed. She came to me one day to ask me to reach out to the Light Beings for her and to find out what Bentley was doing on the otherside. So, in January 2011, the Light Beings gave me a message for her:

> This has been the first of a strange request from our dear Dinah and we feel proud to perform this task, as her little sidekick and your heart animal, dearest Marla, is a character of time. Though not human, his knowledge of the ages is strong and his small heart and soul beats to the drums of the saviors of time. He is the joy of the worlds. Yes, yes, he is as you both have described crotchety, that small devilish being. He makes all of us smile. The guides are continually shaking their heads, yet they too realize the potential of such a spirit. His life with you has been a gracious undertaking on your part, for you have given and suffered greatly as well for the little one. He has always felt the kindness and the love (in the way of animals, which is different than the way of humans) and that should be such a joy to you at this juncture of your

human life. You will see his kind soul again and he will be able to update you on all his travels through all his lives – not just the time spent with you. He goes into great depth of his life as a tree. Yes, how strange that must sound to you, but that is when he first asked to be a bird. He knew that he needed to have the wings to fly. But then he found that as a bird, he could not interact with the new beings in the world that so intrigued him. Humans. Oh, his is a tale of such fun. Yet much sorrow, and pain, and learning has taken place in his life. He is a wise one.

He now acts as a scribe of sorts. Yes, he takes notes for Dinah. And writes his memoirs that seem to tickle her so much. But he sees much more than you or she can ever know. He sees it all and no longer chides the protector horse [Solo] as he did in the core group…he watches and records. He makes sure that all is seen by the horse and by the guides. They sometimes become annoyed at him, but all know his heart is in the right place.

You too, Marla, are looked upon regularly as he cannot protect you, but he writes in his book and will have quite a thing or two to say at your next meeting. We shiver. (And laugh.) He again asks that you know that it was time for him to move on, because of the things that needed doing on another plane.... It was not a time for him to dally in the affairs of humanity. He had things that needed doing.

He misses you greatly but does not cry because he will see you in what seems to him merely a tomorrow.

Be prepared to be bitten. Again we smile.

On With the Show

So with little fanfare, we now move forward to the stories of the animals impacting the world. You will read about the positive attunements that interact with autism around children, dolphins with messages of love, crabs with an agenda, butterflies, llamas, dogs, cats, and even flowers. It's time to pull up a chair and get comfortable, because you are about to realize how all animals impact your world.

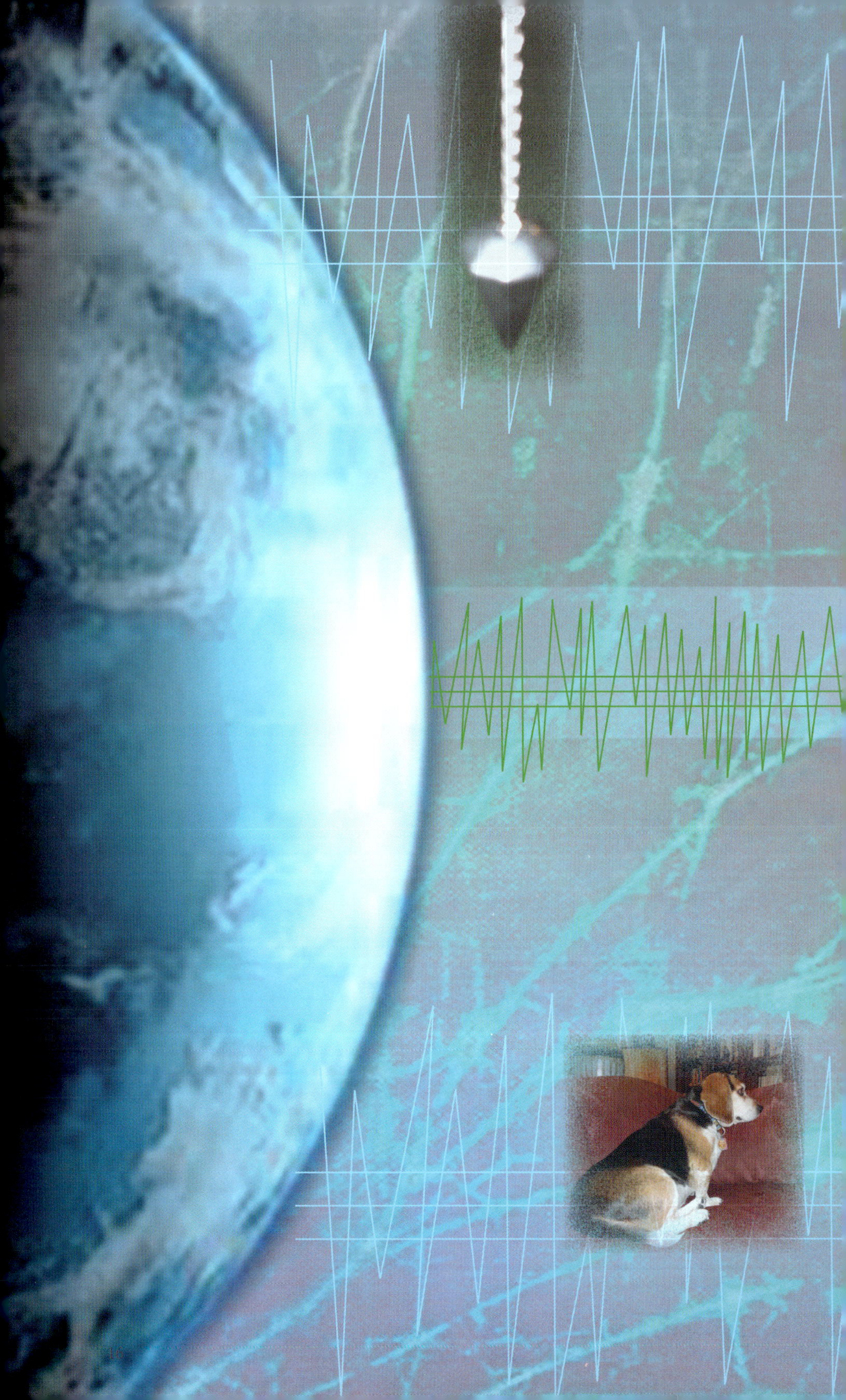

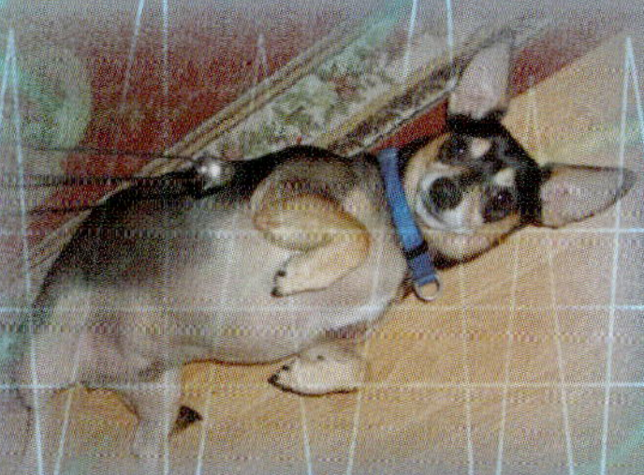

PART ONE

Animal Communications

BASICS

Animal Communications

First Steps and Beyond

"Every seed is awakened, and all animal life."

~Sitting Bull

Though I've explained in detail the steps to becoming an animal communicator in *Psychic Pets: Solving Paranormal Mysteries*, I thought I would start with a bit of explanation for those not having some of the basics included in the first book.

The most important piece of information for anyone wishing to be an animal communicator is that *anyone* can do it! It can be a learned skill. (I say that because some people are born or gifted with other abilities and do not have to start at square one.) Yes, it takes practice and a suspension of thoughts that, in the past, may have said to you that such things were not possible. But once you convince your brain that it can be done, you will find that the most wonderful things will begin to happen. I do suggest you read *Psychic Pets*, because it describes a typical journey and relates the pitfalls that might encumber you during the process, as well as the fantastic highs that come with successes. And above all that, it's fun!

As you will quickly realize when you begin to read this book, what may have begun as a follow-up book to *Psychic Pets: Solving Paranormal Mysteries*, became something much different – not because I wanted it to be that way, but because the animals who were to be involved (and make no mistake, they chose Mary and me before we chose them) went beyond just communicating and on to the impact of communications of all kinds on the human race. Even so, in-depth thinking told us that preparing people to be part of the actual animal communications process that went beyond the normal daily variety would still be important and a necessary part to understanding the impact as a whole – big-picture stuff.

For me, knowing first who my animal guides were was such a fantastically delicious feeling that it would be criminal to deny it to anyone else. I personally feel that anyone who talks to animals without interacting with their animal guides is losing a close and wonderful experience. This, of course, is a preference for this particular book, but you won't be sorry if you identify your animal guide, I promise you that! For once identified, you will see that animal everywhere!

Identify Your Animal Guide

Identifying your animal guide is as simple as meditation or creative visualization. I even use these same techniques in my hypnosis practice. The key is to relax enough to keep your busy mind where you want it - out of the physical world and in the world of animal communications. Although any creative story can take you where you want to go, the following is one I use very successfully when working with others.

Imagine yourself turning to the left from a sandy walkway to a daylight path. Here, you find yourself walking through a field of wild flowers - purple, yellow, white clusters of floral wonder. Their sweet fragrances drift through the air mingled with a warm breeze, and the scents so sweet you can nearly identify each separate bouquet of flowers. The breeze touches your skin gently and it feels like silk being slid across it. The sun is shining brightly overhead - a big orange ball in the sky - and its warmth mixes with the breeze for such a delicious feeling.

As you walk leisurely along, you see something strange on the left edge of the path. It is a roll-top desk. You wonder why it's there and decide to investigate. Carefully now, you roll back the top. There is a large and beautiful cloth-covered book with images of all kinds of animals on the cover of it. There are wild animals and domestic pets, sea creatures, flying birds, and even reptiles and insects. *How interesting,* you think. Gently you open the book, and fan through its pages. You see page after page of animal photos depicting them in their habitats, in people's homes, and in places around the world. What a wonderful book this is.

You notice on the front inside cover there is a slot with a glossy piece of photo paper. At the top you see your name, and beneath your name, there is a blank space for a photo of you with your animal guide. You decide to take this paper with you, for you know that there needs to be a photo on it for it to go into the pages of the book. Now you close the book and the roll-top lid on the desk and continue along the floral path.

Again, you are feeling peace and contentment as you walk along - a beautiful day this is to meet your animal guide… A few steps more and suddenly you notice that your mind has gone to alert. Something has changed in the world. It's as if you have called forth some great and wonderful force. Yes, you have! You continue to walk.

Look down the path. You see a dense forest with trees and brush blocking your eyes from seeing within. As you get closer, you stop, squinting to peer in, but you can only see the greens of foliage and the browns of huge tree trunks.

Still…you can hear something coming from a distance, as the forest is rustling with movement. Something – or someone – is coming. And you know that this is a good thing. Something wonderful is coming. Your animal guide is making its way from the world beyond to be by your side.

You feel excitement, as you stand silently by, waiting for your life to change when this magnificent creature joins you – any moment your new best friend will be here. The rustling increases now…soon…and now, bursting through into the light, your animal guide arrives.

Who do you see?

Photo Credit: Jim O'Rear

Yes, Virginia, There Really is a Santa Paws. . .

I would be surprised now if you did not know who your guide was! Everyone I've worked with has seen some manner of animal come out of the forest – even, as in the case of a sea creature, they bring along their own habitat in a bubble!

So, once you have this information, what do you do with it? Ah, the sixty-four dollar question. If you are anything like I was at the beginning, you doubt it. This is something that I hear from every animal communicator (new and old) who I've talked with. Just like the "I can't do this; it can't happen" kinds of feelings you may have felt prior to looking for your animal guide, afterwards you may feel, "this can't be; am I really hearing the animal?" And then you will answer yourself, "It's my imagination." Oh, so typical. So, not to worry. This is a route well walked, this doubting path.

The doubting path is a strange trail. It can be long. It can be short. It can go away and it can come back. It's always right around the corner it seems. As a trainer and supporter of new communicators, I have a very easy job. The questions are always the same: Am I really hearing? Can this really be true? And the answer is very easy for me. I can wash dishes and still be able to clearly give this answer without strenuous difficulty. I can food shop or dust, and answer! You see, the answer is easy. I just say: Yes. Yes, you are hearing; yes it's true.

Now that's not to say that every single person hears and has the experience in the same exact way. There are many variations to this. Why? This all has to do with the natural functioning of the brain. We all learn in different ways and science has known that for a long time. Some people are visual, some auditory, some kinetic, etc. You will note in the script for finding your animal guide that there were varied kinds of descriptions bringing out all of your senses. That was important because not every person will react to the suggestions in the same way. I am very visual and see things like a movie in my mind. As far as I know, I'm the only one I've heard of who consistently sees a "core group" of animal individuals who help me with communications along with my animal guides. Mary *feels* things; she can feel the sun's warmth or the breeze on her skin. There are people who can hear only; others just know. Therefore, when communicating with animals, there may be some varied perceptions – none are wrong or right. They are what they are. The more you practice, the better you become. It's all about the practice. And a little about the Otherside!

A new communicator may become confused by these varied perceptions, especially if that person can receive signals in several ways using more than one of the five senses (or six if you count the third eye's involvement in all this). So my suggestion then, is to continue to use your script to go down to that path where the animals frequent – or to a path that you find as more relaxing and a better fit for you.

Getting to Know the Animals

Once you finally believe, or suspect, that you know your animal guide for communications, it's a good thing to get to know the animal. When I found mine, I immediately went to the Internet and searched for the meanings for my two animals (the Lemur and the Leopard). You will find that many of the Native American sites have this part of it down pat. They know why animals come to you and what their characteristics are. If you are asking why this particular animal (or animals) come to you, it is because at this time in your life, these animals are what - who! — you need. Some brief meanings for some typical animals are below. (This is, by no means, an exhaustive list of animals or the only interpretation of meanings — you will pick up many other things from your animal that apply only to you and your experience. It is best to research beyond this offering.)

Butterfly

Connection to the otherside, reincarnation of spirit, transformation, learning about light and love by understanding darkness and internal growth.

Cat

Independent thought, visual insight to mysterious places (shadows, otherworld creatures, etc.), an expert in dream recollection and deciphering.

Cougar

The gaining of knowledge, staying the course, self-confidence, leading others, personal power.

Dog

Emotional healing and friendship to humankind, protective and loving nature, movement through troubling times.

Dolphin

Communications of all kinds, learning to trust and to develop positive emotions, one with nature (especially water), balance of the chakras.

Eagle

Great sight into intuitive matters and to see higher truths, wisdom in decision making and hard work (including balancing mind/body/spirit), power and steadfast movement through life.

Elephant

Connected to the memory of the ages, relentless movement through problems and difficulties with life; regal tendencies and the pursuit of higher education in all matters.

Fox

The power of invisibility (hiding in plain sight, silent and learning about all those around you), clever and courageous in thought and observation, fast thinking, gentle heart.

Giraffe

Sight into the future, communicating with those who have been seen as unreachable, strong intuition, remains above the lower emotions.

Horse

Travel (real and astral), freedom, and survival matters, cooperation via connecting to individuals and forging friendships, identifying and overcoming danger.

Humming bird

Patience, survival instincts, love and happiness, healing energy.

Leopard

Stealth, abilities involving communications to individuals unseen (otherworld) and plant life, pursuit of plans until managed and completed.

Lion

Strong and powerful (able to lead with courage), connecting to others (family and those of like minds), excellent use of energy.

Mantis

Stealth, aware of time (stopping it, moving through it, controlling it), building strategies.

Owl

Seeing what is hidden (tearing away masks), keeping secrets, freedom from fear, fully aware of "otherside."

Panther

Guardian force, astral and dream world interpretations, insight to darkness and manipulating that power, rebirth of souls.

Rabbit

Compassion of those around you, quick thought processing and decision making, overpowering fear, strong intuition.

Snake

Messages from beyond, psychic abilities and creative force involving the unknown, elusive and magickal.

Spider

Weaving futures, repairing pasts, understanding fate, shapeshifting, understanding how to manipulate illusive topics.

Tiger

Willpower and strength, quick action-oriented decisions and plans, awareness of the passage of time, energy, and authority.

Turtle

Staying power, abilities in the development of long-term plans, ideas, depending on oneself to overcome violence, earthy.

Wolf

Family life and personal protection, instinctual reactions to change and enemies, dream and meditation abilities, social networking.

As mentioned, these are just a few animal guides that have visited communicators through the ages; there are hundreds (maybe thousands) more. Note that just because you read that your animal represents certain traits, it is probable that the animal is slanted directly to one of your strong needs at the time you begin your journey into communications. It is also possible to have more than one animal guide, to have them change over time (long or short), and to have them laugh at your desires to plant a trait onto them. In other words, be prepared for anything.

And Now, Back to the Path

It is important to get to know your animal intimately once you have discovered his or her identity. There are a few good reasons for this. The obvious one is that you have to have that bond, that connection that allows you to trust what your animal is telling or showing you. But there is a more important reason beyond just you and your connection. Your animal will, in the beginning, show you how you will understand how other animals feel. You will learn how to discern where on or in an animal's body there might be physical ailments or special feelings that need to be communicated.

Case in point (and discussed in *Psychic Pets*): Once I learned how to become my animal, I was able to describe an injury for a horse I was talking to, by recognizing the bodily functions of animals. My left forearm hurt, which indicated that the horse was feeling injury to his left front leg at the lower portion near his hoof. In other words, your animal guide will teach you to shapeshift. I am sure some of you may be rolling your eyes at this statement, but if you research varied earth religions, you will find that one way to shapeshift is to become your animal. How does one do such a thing?

First you must suspend your disbelief (again) that you can't do it. These lessons are all mind things and it has been proven that the mind is a powerful part of you. It can cure you, make you sick, stir you up, or slow you down. It can also become something else.

Once you identify your animal, then there are steps to follow that are the same regardless of the animal, though the terrain may change from one to another. You wouldn't have a dolphin interacting with you from a forest; you'd want to see the water. I must say, though, that at first identification, I always use the forest and all manner of animals (land, sea, or air) have come to me from that place. But that is only a flash of time. You then begin to personalize your own environment and training ground.

The Steps

Once your animal guide has greeted you at the forest edge and you know who your animal is, now move forward into the woods with him or her at your side. (If you are with the dolphin, instead of seeing the forest, see the ocean; if you have a bird, see the sky; if a camel, see a desert. Change your environment according to your animal.)

After you are inside the environment you have chosen, stop and look at your animal very closely. Notice everything you can physically about this animal. See the eyes and how they study you. Look at the skin/coat/covering of his or her body. If you receive impressions from the animal, listen carefully – and be sure to have fun! (I can still remember my daughter's first time doing this and finding that though she had a panther at her side, there was also a cartoon snake "running" fast alongside yelling, "Ooh, ooh, ooh!" She was so confused and taken aback. I laughed hysterically, for I could see him, too! You never know what will happen! But I've never heard of anyone who has had anything but a wonderful experience doing this. I do wonder if having a T-Rex would scare me though…)

Now stand next to your animal facing the same direction and looking forward. Breathe in and out slowly, and then sidestep (with your mind) into the animal. Look out from the animal's eyes and see a new point of view. Take note specifically of the things you see and how differently you now see them.

Begin to move as the animal, walk, run, fly–whatever the method of movement for that animal is. Feel the muscles, or the wings, and notice the environment rushing by. Consider the way the animal handles his or her environment. Understand the environment.

As you move in varied ways within the animal, converse with him or her. Hear their comments or advice. Allow them to teach you how each body part feels. Be the student.

Once completed, gently step back out of your animal and into your own body (which has always been there, but in a waiting mode). You are back in *your* environment with your animal and you can exit the exercise.

This exercise should be done in silence and it will take you roughly fifteen minutes – however, you can stay within the animal for as long as you choose. Too, anytime you feel stressed or out of sorts, this silent exercise will act as a meditation as you allow your animal guide to relax you and to put you back into your highest state, for this is a friend for life.

Once you have met your animal and understand the shapeshifting efforts, you are now ready to communicate with animals. Use the same method to find the animal you need to talk with as you did going to meet your own animal. In almost all cases, your animal guide will be there to escort you to your "client." As you get better and better, your guide will just tag along as you immediately touch minds with those animals around you. Eventually, you may not need the book in the roll-top desk at the beginning of the path to have a photo of your animal. These things are merely to get your mind relaxed enough to visualize along the way. As you become adept at it, the animal may come to you before you can even call for it. Anyone can do it; it just takes lots and lots of practice. And faith in yourself.

CAUTION: NEVER SHAPESHIFT INTO AN ANIMAL FOR A NEGATIVE PURPOSE. THIS THEN PUTS YOU INTO THE CYCLE OF "WHAT GOES AROUND COMES AROUND." YOU ARE INVITING NOT ONLY BAD KARMA, BUT DANGER AWAITS ANYONE USING NEGATIVITY TO MANIPULATE THIS OR THE "OTHERSIDE."

Learning to communicate with animals, even if in a very small way, allows you the freedom to understand where once there was no understanding. You will see the animals in a new perspective and not as merely other species situated beneath humankind. They will come to the forefront and your own lessons in world living will become clear.

The Pendulum and Animal Healing

"Healing is taking place the moment you are intentionally sending unconditional feelings of love [to your animal]."

~Elizabeth Whiter
Author, *The Animal Healer*

As you move forward in the larger-than-life efforts and intricate task of immersing yourself in the lives of the animals around you, you may find that a friend will come in handy. Now, I don't mean the traditional friend—of the human or even animal variety. Rather, it behooves you to become intimately involved and knowing of your "high spirit committee" (the group of guides in the spirit realm who watch over you) and of course your animal guides. Many people have done this, and though it sounds like a task fraught with difficulties, you will find that it can be very easy. That's where the pendulum can be helpful.

Intent

Where to begin... *Intent* is the word of the day when it comes to understanding the world beyond and how your soul and spirit are connected to it. Once you've truly decided to become one with the source (or God as some people call that one great entity), then you will find that intent is the most important part of enhancing spirit, fixing the soul, and healing self or others. In reality, the practice of intent is little more than your religious ideals or as in believing strongly in yourself.

Take the religion of Wicca for an example. An aspect of this religion can (if so desired) depend upon spells and chants to bring one closer to nature. What happens if a Wiccan person practices a spell but has no belief that it will work? Nothing happens, i.e., if the intent of the spell is not true and strong, then there is no outcome and no connection. Varied Native American rituals are also dependant upon intent. Even the traditional religions of Christianity, Judaism, and others call on prayers for assistance to the soul with forceful intent. How many times have you heard that there is power in prayer? There is. Why? Because the intent to heal is in the design of each individual sending out a pertinent message into the universe. There is intent.

So, once you realize and understand that there must be an intent behind any communication involving your own personal spirit, it is likely that you will begin to see the world in a less muted way. It only takes one act of intent for your spirit to recognize the familiar pangs of "knowing." If you focus on your intent as you begin using a pendulum, the results you find will be truly amazing. If this becomes just an exercise in frivolity, then that too will show you something. (It will show you that your lack of intent can bring on skewed answers that feel like lies.) The intent must be managed when using a pendulum.

Pendulum History

Using a pendulum for scrying to find answers to someone's most important questions is not new. In fact, people have been using varied forms of the pendulum for centuries. There have been reports of usage that have been brought to light in ancient Egyptian lore as well as during the rise of the Roman Empire. There are those who say that this technique was available even before history was ever recorded.

Varied materials have also been used for pendulum scrying. For example, it has long been known that using scrying rods (which are linked to pendulums by many) to find water has been successful. There are people today finding wells using a pendulum or metal rods. Women have been known to use pendulums to determine the sex of unborn children with very simple homemade pendulums made from a pencil with an eraser at its tip, a needle, and thread. (Thread the needle and then stick it into the eraser of the pencil. You now have a dangling pencil with a thread "chain." This reacts the same way as a crystal and chain.) Personally, in a pinch and as I often mention, I've used a necklace, holding the chain as the gem or charm at the end of the chain acted as a pendulum.

What Can You Use As a Pendulum?

When I first began using pendulums, I was very excited to find that there were millions (it would seem) available to choose from in stores and online. I found one in a lovely spirit shop in Portland, Oregon, and it was as though it called to me personally—a distinct feeling. I used that pendulum for years, until one day the worst happened. I went looking for it and it was gone from the very specific location I'd kept it for nearly five years. I did everything I could think of to find it, including retracing my steps from the prior day since that had been the last time I'd seen it. I even used rods to scry for it. No luck. I'd felt sure that I must have lost it at my doctor's office but no one there had seen it. Sadly, I knew it was gone. So the search was on for a new one.

I started looking for another one that spoke to my soul, but nothing held that same connection for me as the first one and I was very disgruntled about it. I refused to buy something that I could not have that personal connection with. I began using a little homemade one that I'd been given in an animal communications class that was made by threading yarn through a nut from a tool kit. If I wasn't going to have a connection, it didn't matter really what I used for my link to my guides, and the Otherside was so strong for me at that time, I could use a telephone receiver on a spiral cord and it would swing just like a pendulum. (Seriously, I'm not lying.)

All this to say that eventually, I saw a necklace in an *Avon* magazine that was a teardrop contraption on a fine chain. It was lovely and I began using that for the longest time. Jewelry works well.

As an aside, out of the clear blue, I walked into my bathroom one morning after I'd found the Avon necklace to use and there was my original pendulum of old lying in the middle of the floor. The small bag that held it was not with it, but somehow that didn't register. Gone a year, and now back. In plain sight. (Spirits do have a sense of humor – I suppose I was meant to find that a multitude of items work so that I could tell this to you!)

How to Use Your Pendulum

For those of you who have skipped around in this book, hoping to find immediate gratification, you can learn to use the pendulum here. I'd like to also note now, that the pendulum is a training tool only. Once you become good at both asking and receiving answers, you should find your own pendulum—one that "speaks" to you. It need not be crystal. It just needs to draw you. You will know it when you see it – or feel it. It will have a sensation to it that will be undeniable. Whether it is homemade or purchased from a new age store or convention, it will call out to you. Some people have mentioned feeling an electric sensation when touching their new pendulum; others, whispy air, and still others just have a "knowing."

Using the pendulum for dowsing and healing practices can provide a specific method of divination (healing) and provide answers to questions by the movement of the arms giving *yes, no,* and *uncertain* replies to questions asked. This belief and intent is practiced by almost all animal communicators I have interviewed or have had contact with. It is also one that I subscribe to, since I have personally seen the results by a psychic/ghost investigator and have had results myself via animal communications and energy experiments on my dog, Rufus, who was very ill with congenital heart disease. (I was able to relieve his discomfort somewhat for about two years before his death and in my communications with him, he often thanked me for it.) Other psychic mediums I've seen using the pendulum believe that they are talking directly to a spirit involved specifically with the healing.

To begin with, I find it necessary to learn to use the pendulum in all its varied ways prior to working with a healing. So speaking to the universe through it becomes the first order of the day. The pendulum should be considered as the same kind of tool as the Ouija board – therefore it can be dangerous if the appropriate protections are not put into place. *Why?* you may ask. That answer is simple. No one knows for sure who exactly (or what, if you care to listen to some rhetoric) is answering the questions you are asking. Could it be the animal you are talking to? Is it the guides who are watching over you? What about angels? God? Lower creatures? Or worse? My answer to that is...well, it could be. Therefore, with my particular ritual I make the selection as to the who or what when using my pendulum. It wasn't always that way – back in the days when I didn't know better. But now, I've made it clear that only my guides who have my highest good at heart be allowed to talk through my pendulum. I ask them to be both gatekeepers and translators. If they advise me that it's safe to talk directly to a specific "entity" (could be animal, ghost, angel, Light Being, etc.), *then* I do; if not, they translate answers for me. That way I know I'm safe. I always put up protection (by visualizing the white light around me and sending a protection prayer to the Archangel Michael) for good measure. (No sense taking chances.)

Tip:

A pendulum can be used to clear Oracle or Tarot cards. I've also heard that some use sage, say prayers over them, leave them under the full moon, or conduct a specific ritual or spell. There are many ways to do this.

The Basic Mechanics

Once you have chosen what the modality of your pendulum will be, it is quite simple to use it.

- The first step, as mentioned prior, is always to say a protective prayer or conduct a ritual that will protect you from negative forces. For healing this is so incredibly important because you do not want to make matters worse by inviting in negativity or more sickness. It is at this time that you might like to put up some parameters regarding the "who" when it comes to finding out who is giving you the answers. I ask that all communications come through my guides. But there are times when I will allow other communications, and these I inform must be "okayed" by the gatekeeper guide. This is the guide who keeps a strong eye on everything and everyone who gets through the "door" to talk with you. Both are safe ways to go.

- Next you will want to find out how your pendulum would like to work. As you hold the chain or cord between your thumb and forefinger of your dominate hand, ask it to show you what a yes looks like. In many cases it will swing clockwise or may swing up and down (as though nodding in the affirmative). Then ask it to show you a no, followed by a request to see unknown or any other kind of answer you may like. There are also charts that you can make (or download for free online) that allow the pendulum to swing to letters and numbers. (My patience wears thin for this method - it takes a long time to spell things out, but it can be done.)

Then, you are on your way. I find it best to record responses and then compare them to situations, because, as mentioned, variables change which change the environment around the pendulum's answers.

Healing

One wonderful way to assist a pet with healing – and if it seems that you do not have a *formal* "healing knack" – is to use the pendulum to ease illness. This works particularly well with issues involving pain or if it appears that the chakras are blocked, keeping normal healing from making its way to the animal. (I'm afraid that studying the chakras is a whole different line of work, so please look more deeply into that subject.)

Many people use Reiki, traditional veterinarian medicine, and other holistic remedies to treat animal illnesses, and the pendulum can strengthen these by providing an energy conduit that is usually unseen or may be unavailable to these other methods. In the same manner that you use the pendulum to receive *yes* and *no* answers to questions, you can use its "swing" to raise vibration levels, allowing other healing measures a better grip.

The good thing about using the pendulum for healing is that it seems to respond much better and with visible accuracy that is

Here Louie gets some pendulum healing for itchy skin. This particular pendulum has a clasp that you open and fill with something important. In this case, there are healing stones inside (research gemstones to find which stones will work best for your healing efforts). Other items can be placed inside as well – a calming agent like lavender for example.

not often available when using it for the answering of questions. When using it for questions (and I definitely do use it for that in animal communications or as a clarifier when reading cards), there can be discrepancies because the answer for that moment reflects only the solution that is available at that very second with no other variables included or changes that might occur due to the knowledge received. In other words, mixed messages or confusing answers can abound. For healing though, this is a "made in heaven" kind of proposition, because the guides and angels working with healing have no variables to contend with on the human front. Healing is healing is healing. The pendulum swings! (This by the way can be used on people as well. However, I've found that unless the person has an inkling of believing, they can affect the outcomes negatively just by choosing to disbelieve sometimes. Animals don't do this! I should mention, too, that there are many who believe that the recipient *does not* have to believe for successful results. This is an unknown for me.)

In my experience, the pendulum, when held over the area of concern, will swing clockwise in wide circles, slowly moving towards smaller circles, until it finally stops. This means that the amount of energy available at the time for the particular place on the animal's body has been absorbed by the animal. It is then wise to follow the same patterns all over the animal's body. Providing energy for the entire body in addition to just the area of concern allows the animal to have a pool of energy that can be gathered from other areas, if needed then or at a later time.

Does this cure the malady? Not usually – at least for me. I have known healers using this method that produce full cures, though. If one has a very real healing skill, then this method can indeed heal or provide strong results, but for most individuals, I find the process is just a strengthening procedure that gives the animal a better fighting chance at healing when combined with other methods. Understand, too, that the energy is not only coming from the universe, it is also coming from you.

Don't forget to thank those who have helped you with the healing: the guides, Angels, etc.

Issues to Contend With When Doing an Animal Healing With a Pendulum

Sometimes the animal will think you are playing (especially if you are healing a cat) and they will bat at the pendulum. This takes your concentration (your intent) away from you and means starting over.

Getting your animal to be still for a specified amount of time could be a trial. It's best to try your hand when you can sneak up on the animal as it sleeps.

You need to *know* that you are helping, otherwise you're not. Many people want to see instant results – this is not a usual thing. The animal, once absorbing the energy, then has to allow the body to pull that energy to where it is needed and in a way that it can be used. It's not usually done instantaneously.

For animals that are very, very ill, it may even seem like it is not working at all. Trust me. It helps. The energy may not be enough to put the illness at bay ever, but any tiny bit of energy you provide is a help. If nothing else is clear to you, it sends your much-needed love into your animal – and that in itself is powerful.

If your animal shows discomfort and tries to get away from your healing, this is because he has had enough or cannot take more. Respect your pet's wishes and stop right away.

Tip:

The biggest deterrent to healing practices for your animal with the pendulum might be listening to anyone who leads you to believe that it can't be done. The slightest doubt can hurt your animal. (Always remember: Who cares what other humans think?)

Pet Grieving

"Grief makes one hour ten."

~William Shakespeare

Most people have experienced the loss of their beloved pet and have gone through all the stages of grief, then the healing process. Recently in the Gasparo home, Mary went through that horrible experience of that loss:

BJ, the first born of the household, left a void in our lives. We were blessed to have her with us for eighteen years. As hard as her passing hit us, the impact was harder for the rest of our feline clan. One in particular, Gracie, was consumed by the loss. She grew up with BJ. After, BJ's passing, Gracie laid for days in the last place BJ rested. I couldn't get her to eat or drink. She refused to leave that very spot. Eventually, nature took over and she began to eat and drink water again. Gracie stayed distant from the others. She often would go to the basement late at night and cry out. Her wails were heard all throughout the house. This went on for months and was very difficult to hear, because I knew the pain/grief she was going through. My heart was broken, because I knew this wasn't something a cat treat could fix, but only time.

At first, I thought these behaviors were just unique because Gracie had grown up with BJ. Then, less than nine months later, Gracie lost her own battle with kidney disease. She was only seven years old. During her life, Gracie had two litters of kittens. I kept a female from each litter – eldest daughter, Daisy, and her youngest, Dori.

This part of the story focuses on Dori. Gracie and Dori became kitty friends rather than mother and daughter. They played together constantly. *Tag, You're It!* seemed to be their favorite. It was a nightly ritual. When Gracie passed, Dori was the one to grieve the most. She didn't go to the basement like Gracie did when we lost BJ, but she stopped eating. The sparkle in her pretty blue eyes seem to fade. She loved to play and chase, but she'd lost her playmate. For weeks, she went from one to the other (I have four cats) to generate play and chase time, but the others were not inclined. I watched her getting thinner day by day. Then I thought to myself, if she doesn't get over this, I will lose *her* too. So, I had to think of something to snap her out of it.

I bought a toy, which consisted of a small pole, ball, feathers, and a bell on the end. What cat doesn't like feathers? I began to sit in the middle of the living room floor and started to swing this new toy around on the floor. One by one, they all started playing…TOGETHER, waiting turns for the other to have a swat at it. They would get so riled up after the playing began, Dori was included into the fun. It took awhile, but the furry family is exactly that, a family.

As for pet grieving, I fully believe they do grieve for the loss of another. They may not all cry, but they react. Don't ever dismiss the notion that animals do not feel, because I know firsthand they do. Most of the time, like Gracie's reactions to the loss of BJ, the pain of loss will eventually heal. However, a little intervention sometimes is needed, as in the case of Dori.

There are five stages of grief recognized by our human medical fields. They are:

1. Denial and Isolation
2. Anger
3. Bargaining
4. Depression
5. Acceptance

All have stressful consequences for anyone going through any one of them. Now, if we as humans have these stages, and we believe that animals are living, breathing, thinking, feeling creatures, does it not stand to reason that they too may go through these stages? Their reactions might not be ones we see or understand, but some they communicate to us very clearly, as mentioned earlier.

So, how might an animal react if experiencing these stages? And this applies to an animal grieving the loss of a human, as well.

Be open to varied ideas to help your pet. A toy worked for Mary this time, but others have mentioned different things you can try. Make sure that your pet is well hydrated and is eating. Exercise the animal – walking or playing, just some kind of movement that keeps him or her healthy while recovering from loss. Adequate rest is important, too.

It's good to know that these stages may not come in order! Just when you think you might be out of the woods with your

Stage	Behavior
Denial and Isolation	The pet may keep searching the home. Pacing. Hiding. The animal may seem to see things that are not there (in your frame of mind) – they may see spirits of the deceased. This may or may not spook them.
Anger	Difficult to tell – some become destructive, but most just stay moody for a period of time. They miss their connections but they do understand that "life goes on." Angry behaviors can also be aimed at family members, especially those the animal may have felt (or feels) he or she is superior to.
Bargaining	Since this is totally an internal thing, there's no way you can know unless you practice animal communications skills or hire a communicator. We do know that animals are more understanding of their place in our world and therefore this step might be skipped entirely, just because they are smarter than we are. There can be bargaining if a walk-in is involved (see *Pet Psychics: Solving Paranormal Mysteries*). This becomes complex and may require an entire new way of thinking about reincarnation, your own loss and desires, and the connection between you and your animal.
Depression	This could last for a long time – or not. It will be different for each animal. A strong support system will help minimize this or shorten the duration. But make no mistake, the depression is as serous as yours. Caution: Do remember how you yourself grieve. If those around you try to show the "sunny side," that triggers the need to hide away from "understanding eyes." This could happen as well with animals.
Acceptance	One day your animal will seem to be the "same ole pal" as before. This mirrors human reactions. Do not be surprised if your animal oscillates back and forth between many of these patterns for this is common (in animals as well as humans).

pet, an angry/moody stage might reappear and you're off again trying to help him or her cope!

This is by no means a do-all/end-all look into grieving for animals. You will find that the personalities of animals are so varied (just as people) that trying to predict behaviors or react to them may be different all the time. I suppose that the old cliché of "Love conquers all" applies here. Love and time.

Just know that your furry, feathered, or otherwise covered friends depend on your support in a time of grief. They are always there for you and there are times you need to be there for them.

Don't Grieve, Believe!

Spend Time

This is the best way to help them along. Let them know that though they may feel alone, they are not. You will not abandoned them. Do not be surprised if you believe that the depression and loss stages have been navigated and that it's all over. Grief can return even months later – again, just as in people.

Recognize Their Needs Right Away

Is the cat box dirty? Clean it. Do the dogs need a walk – just for the fun of it? Do it. Though it may seem like nothing to you because it may be daily behavior, to them, they are appreciated. Just like we hate daily grind things when we are grieving, this must be so for animals, too. Doncha think? But sometimes those daily things are what keep us going.

Arrange Activity

New things to do, new patterns to emulate. Just like the cat toy that brought a family together, so can a family activity, a special new ritual... Put on your thinking caps!

Beware

Do not judge your animal based on your own feelings. Just because you may be "over" a loss, does not mean that your pet has recovered. This is similar to people not understanding people and how others are different than themselves. (Doesn't it anger you when others think that you should "just get over it"?)

And believe it or not, once you start helping your other animals get over the grieving process, you will have had help with your own grief!

Remembering and People Grief

The following is a story about a very special, sweet girl that came into both Mary's life and mine at different times, but in the same place – at a paranormal group meeting.

Zira was the "hostess" for the Chester County Paranormal Research Society (CCPRS), where we both garnered a lifetime of training and good friends.

This wondrous dog was present at every meeting the group had and stuck it out no matter how far into the night our rantings and ravings lasted. We loved her as our own and considered her part of the proceedings. It was never in anyone's mind that she would have to leave us one day.

Zira's life partner was Mark Sarro, head of the CCPRS group and he'd spend many years loving her and communicating with her. But grief takes no heroes, and when the time came for Zira to move on to the otherside, his heart broke just as those of us who have found ourselves in similar situations at one time or another.

One way to acknowledge grief of the human variety is by remembering the beginnings, the middles, and the endings, both good and bad – writing them down and sorting through the memories. A tribute to love and life lost. So here for you is the grieving story of Zira written from Mark Sarro's heart.

Zira looking out over Little Round Top in Gettysburg, Pennsylvania, at her last hurrah. *Photo Credit: Mark Sarro*

Zira, My Best Friend

When I was offered an opportunity to write about the time I had with my dog, Zira, and the impact that she had on my life, I had to stop and really think about whether or not I could put into words what she truly meant to me. She was my best friend, my child, and my companion. We had shared a special bond that I think that those who don't own pets can ever truly understand or relate to just how much of an impact they make on our lives.

The beginning. . .

My brother, Nick, had been working in construction/ contracting most of his life. He was out on a work site in New Jersey and came across Zira and her three sisters with their momma. The story goes that the owner of the mother dog had declared that they were getting rid of the puppies by any means; which to my brother's interpretation meant that death was included as a solution for getting rid of them. He immediately panicked and decided to take all four of the puppies. Without question or thinking, he took them and decided that he would figure out what to do with them once he got them to a safe place.

On that fateful day when Zira would come and be a part of my life, I had no intentions of getting a dog or anything of the sort. I had just brought Smuckers, a calico cat, into my home just a few months prior and was starting to think that that was something I did not really think through. I'd never had a cat! My family was a dog family, so bringing in a cat and having it as a pet seemed very foreign to me. My brother arrived at my house unannounced with two of his little guests. He came to the door with pups in hand and brought them inside. They were so tiny! He explained the story and nature of how they came to be in his possession. They were too young to be taken away from their mother; only about 3 weeks old. He put them down on the floor and I began clapping my hands to get their attention; immediately Zira came running over to me and right into my arms. I knew right there and then that she would be mine and that there was no going back.

I would find out that she was the runt of the litter and the smallest of the four pups. She had a grayish coloring to her coat, baby blue eyes, a white patch on her paw, and one on her chest. Her ears were floppy, but she had the face and look of a German Shepherd and we speculated that she was a Shepherd-Rottweiler mix. It wasn't until much later in her life that I would figure out that she was a mix of Shepherd and Border Collie.

Those first few weeks and months were the toughest. She didn't know how to do anything! I had to help her learn how to walk and keep her in my sights nearly 24-7. I would go everywhere with her and at times would even take her to work. I did not want to leave her alone. Since she was such a young pup, I knew it was going to be a while before she would successfully be housebroken, so it became an exercise in patience. I didn't mind; she was so cute and I knew that I already loved her and would care for her no matter what.

As soon as she became old enough, I started to work with her on her obedience training and teaching her the signature "tricks" that my family taught all of our dogs growing up: sit, sit pretty, give me your paw, lay down, crawl, roll over, jump, play dead, etc. She was a smart dog and very quickly started to learn these tricks. Again, it was an exercise in patience and determination, but I would not be discouraged.

Over the course of that first year she had her most difficult challenge developmentally as far as her personality and training. She began to develop neurotic and insecure tendencies. When I came home and would call her name, she would run and hide, pee on the floor, or the worst would be that she would try to get out of the house past you and take off running. I chalked the behavior up to the side effects of a pup that was taken away from its mother too early and that these were things that would hopefully pass with time. I would come to learn that after moving out with my roommate at the time and into a new place, during our time together, he would often take out his frustration with me on the dog. He apparently did not like her, and when he was mad at me, the dog took the brunt of the abuse. If I had known that it was going on and that he was doing those things, I would have certainly put a stop to it and probably would have beaten the crap out of him too! He definitely deserved it.

The bond was strong and growing fast between me and Zira. She slept by my side – either on the floor by the bed, and as soon as she got big enough, she would be up on the bed sleeping right next to me. As she got older, her eyes changed colors to a deep rich brown and her coat lightened up to a tanner lighter coat, but the white spots on her chest and paws remained.

I had moved into a house with a new round of roommates. They fully understood that I had a young dog and it did not prove to be any problem. Zira was still shaking off the neurotic behavior and tendencies that she picked up while being at the prior house. It would make it challenging for my new roommates at the time, because all they would want to do was pet her or be kind to her and she would still hide and be shy and scared.

Over the course of her first year and into her second, there were some challenges, but she slowly began to shake the early neurotic behavior as the environment that we now lived in was a happier and safer place. My roommates loved her and would never think of doing anything to hurt her – especially as a means to take out their frustrations with me. Unfortunately, the tensions in the house would slowly grow. We had a difference in opinion in living habits and bill paying. I found it difficult to live in a house with smokers, when I didn't smoke and had allergies and asthma to go with it. Well, the year with those roommates ended on a sour note and the friendship could not be salvaged. Several months before the end of our lease together, I had to remove Zira from the house. She went and stayed with my sister in Delaware for a month while I tried to make arrangements to get her somewhere more long-term. It was at this time that

I began dating Kat, who would eventually become my wife. Thankfully, due to her kind and wonderful family, we were able to send Zira on doggie summer camp and she went out to stay with Kat's mom on the farm in Flat Rock, Indiana. The meeting of Zira with Luke (the Giant Schnauzer) and Belly (some type of hunting/hound dog) would not go so well at first. Zira was an only child (besides the cat) and so her adjustment to living with two big dogs was very challenging at first.

We left her on the farm for the summer, and that day that we left, I was so terribly heartbroken. I had never been away from her for any length of time. I remember clearly as we pulled down the driveway away from the farm that she came running after us, stopped, and just watched as we pulled away. It made me heartsick and sad to know that things had gotten so bad with my roommates that I had to leave her in a place nearly 600 miles from where I was living. I knew that it would be a few months at the least before I would get to see her again.

When we returned, her personality had gone through a change. The neurotic and scared behavior had passed. She was showing signs of a young and confident dog. She came running up to me and right into my arms. She was so happy to see me, and I her! I immediately noticed that she had grown several dread locks in the fur on her ears. I would come to find out that she had become quite a river dog and would go off exploring and swimming on a nearly daily basis. She looked truly happy and I was so grateful to be reunited with her!

Kat, Zira, and I returned to Philadelphia once again as our happy family was reunited. Until when we got married, Kat and I would continue to live in South Philly for five years with various roommates. At that point, it was decided that it was time for us to be on our own and *no more* roommates. Zira was going on four years old now and had changed dramatically in her personality and demeanor. It was all for the better, because she was now the happy confident dog that I had always wanted her to be.

The bond between her and I grew stronger and stronger over the years. We were nearly inseparable. There was no greater feeling then coming home from work and to have her run and greet me at the door, tail wagging, panting, and barking... Generally excited! "Daddy's Home!" is how it felt from her. I loved my licky pup (one of her many nicknames that would come over the years). During our run with roommates, we had

Tim and Jim living with us who really loved Zira and were very kind to her. Tim, being one of my best friends and the man responsible for getting Kat and I together, is a very creative and funny person. We were sitting around late on a Friday night, bored with nothing to do. I chimed up, "Let's make a parody horror movie!"

"Huh," Tim replied.

"Yeah, I have a camcorder. We can make a short film and edit it all as we go!"

So, it began that night… We proceeded to make *Smuckers of the Undead,* a short horror spoof that would feature Smuckers the cat as the main monster, Kat as the helpless victim, and Zira as Detective Poopy Pup Peterson, the hero that would come and save the day. We made the film and it turned out to be one of the funniest short films I had ever made! It was so ridiculous that you couldn't help but laugh at it! So, from then on Zira was a movie star! (No, not really.) But, it's one of many examples of how Zira was so much a part of my life. She was one of the friends that would come and hang out with everybody. She had to be in the room where everyone was and would be happy just to lie on the floor and watch everyone or even climb up on the couch and lay next to someone from time to time. Yes, she was a sixty-five-pound dog who thought she was a lap dog; and make no mistake about it, if she wanted to climb up into your lap, she would!

The emotional bond between Zira and me was very strong at this point. There seemed to be unspoken communication between us. She knew when I was sad and upset, and I too knew her emotional state. Without fail, when I needed someone most, she would come and be there by my side, putting her head into my lap and giving me little licks of comfort on my arm. I would look down at her and see her big brown eyes staring back up at me, and the sense I got over and over again was "Daddy, you'll be okay."

She was my protector, my friend, and my child. The emotional bond was strong and we looked out for each other constantly. There was never any doubt as to how she was feeling and what she was up to.

Zira was going on 5 years old; she had been fixed at about age 2 and finally started putting on some weight, having been a very skinny puppy in her first year or two. We decided to move the family (Kat, Smuckers, Zira, and myself) out of the city and into West Chester, Pennsylvania. It was time for a change and the city had been wearing on all of us for quite some time.

As we settled into our new home and life in West Chester, we soon began to realize that the house itself was quite haunted! The first signs were in the way that both Zira and Smuckers would react to things in the house. Oftentimes, I would go and stand by the back door and watch Zira in the yard. We had a six-foot privacy fence that surrounded the yard and Zira would sit in the far back right corner, turned as though she was in a "Time out," but her tail would be wagging and she would be looking up as though she was interacting with someone. Smuckers would exhibit the same behavior too; these were the early signs that something was going on with the house. It was not too long after those types of incidents began to occur that both Kat and I would begin to experience the spirits of our home. Zira seemed to be very sensitive to it, and, at times, would run and take cover by my side, waiting for the all clear so she could go about her day (playing, eating, sleeping…).

It was not too long after moving into the house that Kat and I decided to start investigating the phenomena and to try to understand what was going on. We both had had a lifetime of paranormal experiences. I had been an avid enthusiast and had worked several cases over the years as an independent consultant for individuals and families. The bringing of us together into this home in West Chester was the last sign we needed to tell us that it was time for us to start our own paranormal research group. We hoped to help and educate those who have had, and were having, experiences similar to what we'd been dealing with growing up. So, it was there in that house that we began The Chester County Paranormal Research Society and would soon boast a team of members that were very much like us and interested in the research as much as we were. The events of the house also appeared in my book *Ghosts of West Chester*, where I included stories about the house and how Zira and Smuckers interacted with the spirits.

In 2006, we were a few years into the house in West Chester. Life had slowed down a bit for all of us. I think this was when Kat, Zira, and I were at our heaviest. Zira had definitely become a happy plump dog and was probably a little overweight, but we knew she was happy and healthy.

We had begun to use the house as a research study and training ground for the new members of the paranormal group.

Once a month, we would host the members of the group for our meetings and training sessions. Of course, Zira was right there in the middle of it all. The unofficial mascot of the group, she was happy to sit and be a part of what was going on. Because she was so kind and friendly, everyone took to her as well and often would spoil her with the "people food" that was readily available during our meetings. Yes, she was happy to be a part of the group!

On several occasions, during our spirit circle and EVP sessions that we would conduct in the house, she would be lying nearby just outside the circle. These sessions would often begin late at night, so, she was a tired pup and would usually fall asleep while the group worked. I would look at her and say, "Time for bed! Go upstairs and go to bed!" She would perk up, look at me, and slowly rise to her feet to make her way up the stairs; halfway up she would stop and look at me and give me the face that said, "Do I have to?" I would motion again and point up the stairs and then she would slowly continue up; she would pause at the top of the stairs and wait. I knew what she was doing; like any typical kid would do, she would wait and deliberately delay hoping that I would change my mind and let her come back downstairs. "Go to bed!" I would repeat affectionately. Then slowly, *click, clack, click, clack* – the sounds of her nails hit on the hardwood floors as she made her way down the hallway to the bedroom. This was one of our many routines that came over the years (most notably in the house in South Philly). She would try her best to stay up late with everyone and would almost always fall asleep on the floor. Sometimes she would fall asleep with her teddy bear in her mouth – at times while she was chewing/teething on it.

Who Says Cats and Dogs Can't Be Friends?

When Zira first came to live with me, Smuckers had already been in the house for a few months. She was a feisty little calico, and even though she was very affectionate, she wouldn't take crap from anyone – *especially no dog*! Smuckers was easily twice the size of Zira when they first met. Zira, being an inquisitive and playful puppy, would often run up to Smuckers and try to play with her. Well, as you can imagine, the typical reaction from Smuckers was "*WHACK WHACK*" with her paw on the nose of Zira. Zira would turn her tail and run away, but never gave up. Over the years, as Zira grew and became much bigger than Smuckers, she never forgot what it was like when she was a puppy and Smuckers would beat her up! So, Zira had a healthy respect for Smuckers and would not cross her path without her explicit permission. If Smuckers was sitting in the middle of the floor, Zira would calmly wait for Smuckers to move before she would try to pass by her and go into the other room. She knew better then to mess with Smuckers! She was the queen of the house!

In 2008, Kat and I would separate and eventually divorce. It was, by far, one of the most difficult times for me in my entire life. So much had happened in the marriage that hurt the both of us very much. Zira was all too aware of the emotional stress and turmoil that both Kat and I were suffering. She was very much in-tuned with the both of us and over the years was there for Kat just as much as she was there for me.

Zira and I continued on together until her passing in May 2011. In June 2009, she was hit by a car in front of the house, the incident being more traumatizing to me than causing any physical damage to Zira, but I was not taking any chances. I took her immediately to the vet emergency room. Upon their initial examination, they had found things that had required them to perform further testing. The end result was that she had tumors on her spleen, kidneys, and lymph nodes. The doctor suspected cancer as the most probable culprit. Her long-term diagnosis was not good; so every day with Zira, from that moment on, was a day that I was grateful for. She hung in there for almost two years, and it really wasn't until a couple of months or so before her passing that her health really took a turn for the worse.

Zira's Last Hurrah!

Zira had a stroke in March 2011; it was significant enough that she lost use of most of her tail and had paralysis in her back right leg. I knew at this point we were truly on borrowed time, so I had made arrangements to take her with me during the Gettysburg weekend of investigations that my group does yearly. I did not want to run the risk of something happening to her while she was with one of my family members. The hotel that I stayed at was within walking distance of both investigations, so I could easily walk back and check up on her as needed. That Saturday, the team headed out into the battlefield park.

Mark remembers Zira "smiling" in this photo from Little Round Top – a way he will always remember her. *Photo Credit: Mark Sarro*

Zira came along and together we climbed the trails; stood at the top of Little Round Top and climbed to the very top of the observation tower at Culp's Hill. (It was all her doing; I had no intention of taking her up there. She saw the group going up and had to go up, too!) She ended up dragging me all the way up and we passed everyone in our group up the stairs along the way! I knew in my heart that this would be our last adventure together and so I wanted to make the most of it.

The hardest decision a pet owner has to make is the time to say goodbye.

At the beginning of May 2011, Zira's health began to deteriorate daily. It got to the point where I would sit up with her all night and lay on the floor with her on her pillows. She was too weak to get up on the bed and getting up and down the stairs began to be a challenge and treacherous. There were a few nights when I thought she was going to pass in her sleep, but I honestly think that she held on because she knew that I was not ready. I had begun mentally preparing for that day as soon as I knew about the tumors, but now that day was staring me in the face and I had no choice but to face it with her. She told me she was ready and I knew it was her time. In those last remaining days I would lay with her and tell her that it was okay and that I would be okay. On that day, I took her for one last walk and hoped to create one more memory, but the ol' girl was just too weak to go on. We made it a block and a half from the house and I had to carry her back in my arms. I cried the whole way; I knew that it truly was her time. Thankfully, despite the hardships and tough times that Kat and I went through, she got to be there at the end, too. Zira was such a part of Kat's life as Kat was a huge part in her life. Kat loved her and took care of her as though she was her own pup. Kat and I said goodbye to Zira together; I lay with her on the floor and held her in my arms down to the very moment. As hard as it was to do, I wouldn't have wanted it any other way. She was not alone and I could say goodbye.

There isn't a day that I don't think about her and how wonderful she was and grateful I am to have had her in my life. She truly was my best friend and I will love and miss her always. The days and months are now passing by, but the good memories of her grow stronger every day. The times of her sickness were small in comparison to the many good, happy, and healthy years that she had. She will always be in my heart and I know even in spirit, when I need her most, she is still there by my side.

My dog, my Zira, my best friend.

And so we all said goodbye to Zira in our own ways, also happy to have had her with us the short time allotted.

There are usually two kinds of people when it comes to grieving over a lost pet. One goes right out and gets another puppy or kitten (etc.) – not to replace the memory of love lost, but to redirect affection in the hopes that some of the grief will dissipate. Most don't even realize that this is a coping method that relates very deeply to grief. The other group may go for long periods of time – or never – before bringing another animal companion into the home. This is a side to grief that some must go through, for the pain is often so bad that nothing will take it away, and eventually, it becomes a scar tissue, never to be fully repaired to the original state.

Having a deep communication with an animal in any measure puts that relationship on the same level as other relationships in one's life. A very good part of Mark's story, in my opinion, is that he is a paranormal investigator. He has a sense of the world beyond this one and what goes on there. Some of the fear for a loved one is calmed in these situations, knowing that there is another life on the otherside. He can also, should he choose, communicate to the otherside perceiving signs or "knowings," and he recognizes that his dear Zira will always be there for him – even now.

Thus we all look at our grieving techniques, and the grieving of our animals of their loved ones, and we move forward, knowing that one day ...in some place or time... we will all be reunited.

Rock on, Zira.

There is a special place in heaven for those who adopt animals. There's a place for those who donate time and money for their survival. But a very special place is reserved for those who adopt animals that are elderly, sick, or have special needs. Thank you CeCe Wick and Donna Baker.

PART TWO

The Stories

A Sad Time in the Gulf

"Not understanding that you, as humanity, are not in charge of the resources in the way you think you should be has caused much suffering. When will you learn?"

~The Light Beings

As the oil escaped from the bottom of the ocean floor in the Gulf of Mexico in 2010, I was given a specific Channel from the Light Beings to impart here. It is, indeed, time to learn.

Channeled Dedication From the Light Beings During the Gulf Incident – 2010

Mornings come and mornings go, but animals are in our hearts always. Maybe not, you may be thinking, as you are on your way to work or sitting behind your desk, or worse, in a traffic jam. Maybe you are positioned behind a car where a small dog is starring out the back window at you, not barking, not howling, but just looking at you, into your soul, perhaps. Or possibly you have an office where animals reside and spend their day moving from cubicle to cubicle for the occasional treat or pat on the head. Still more of you may be thinking of your own sweet pets at home sleeping in wait for your key in the door or rummaging through the trash bored that you're gone, or worse, in an anxiety attack and destroying your furniture in a desperate attempt to take the fear from his belly because: What if you don't come back?

Ah, the animal and his life. He lives for you and beside you.

But not all animals. Recently, many of them died and many of them screamed out into the night with no one to hear. But we did. And maybe you did, though you didn't understand the ache in your heart or the tear in your eye. Not that you didn't know on some interior level that it was happening. Your television told the tale as the oil was being wiped away from a bird, and small turtles were being taken from the water to be delivered to safe havens. Or when you heard of dolphins moving in the wrong direction, their sonar affected in some ghastly way and fleeing to a place where more oil would assail them. It was difficult to turn a blind eye to this, and many kind and wonderful folks made their way to the Gulf to help and many others sent resources to help with the aid. Others of us just cried in the night, a part of our souls deadened with the knowledge that there were those who would never be on this earth again in that particular way. Yes, it's true that they moved on to a better place, a clean place. But still, their time here was cut short through no fault of their own.

Mankind and those who would fight the good battle would have you believe that it was all man's fault that our animals suffered – and are suffering – so, but that is not the whole story. The blame can of course be shared, but the oil spill did have origins from deep within the earth, and the earth changes are indeed coming to the forefront around the world – and this was yet another sign of that. Yes, this was a terrible accident and not only animal lives were lost, but many good people. And there was enough blame to go around for every one – not just those who were taking the blunt of it. For have you forgotten who uses the oil? Yes. You have forgotten. In the crowd mentality of attacking those in front of you, it has been forgotten why. But now this is neither here nor there. We have come full circle and at the present we focus on the animals – who by the way do not use oil.

The importance of this terrible event is paramount. In reality, this, or something like this, needed to happen, so that humankind could see the damage. Shamefully, the lessons are wasted on those with the most power (where animals are concerned) but the general public has been all too aware and a cry has sprung forth that has lifted voices to the very highest in the universe telling the masses of the love and horror felt at the abuse heaped upon such innocent creatures.

This then is a special dedication to those who cannot help themselves, to those who have passed from this earth with a slick of oil penetrating their lives, and to the people who so valiantly fought and cared for them.

The Elders

Dolphins

"The sea, once it casts its spell, holds one in its net of wonder forever."

~Jacques Yves Cousteau
French Explorer

Art Credit: Eva Sakmar-Sullivan

As we worked with this book, we were continually taken to people who have had life-altering stories involving dolphins. These were creatures that made such a positive impact on the world that we could not consider leaving them out. So we begin by telling you about our first inkling that dolphins had a place in the world of animals creating wonder and hope for humans.

A Vision Worth Swimming For

Dinah met Eva Sakmar-Sullivan at a mind/body/spirit convention in Pennsylvania. Eva was a Visionary Painter/Environmentalist and had huge paintings displayed, most showing the beauty of dolphins as they intermixed with angels. Her work was physically three-dimensional, with the paint spread in a fashion that indeed had to come from her own interaction with godly beings. She has spent her time in oceans (Atlantic and Pacific) swimming with dolphins and sea turtles.

During a sabbatical to create her beautiful artwork, she found that her visionary skills had propelled her towards something different. She was introduced to dolphins through an article that she'd read. She'd always been drawn to dolphins in a small way, but now a compulsion took over.

This compulsion that Eva explains is evident in many animal communication stories as well as in the mind/body/spirit world as a whole. Suddenly, it would seem, out of the blue, a tremendous desire will be put upon one's heart and nothing can stop forward movement towards satisfying it. What Eva found from the article was that people were swimming with dolphins and that she, too, could do this. The desire was set. She knew somehow that this was a life-altering step that would affect her both personally and via her artwork. Calling the place offering the trip to swim, she signed up, paid the fee, and let her excitement boom!

But then, as she says: "Reality set in."

Eva didn't know how to swim, having a horrible fear of water because she'd nearly drowned in a bathtub as a child. (Now for those mind/body/spirit people in this reading crowd: Is Eva learning now one of her life's lessons about water or is she preparing to offer her heart to the world with her paintings of these wonderful sea beings? We think both.) Anyway, despite the fears, the desire kept her on track. She was going to swim with the dolphins.

First, then, she knew she needed to learn to swim prior to taking the trip, so she trained for several months at her local

YMCA. But depending on her own inventiveness, she found that swimming was not to be a strong suit. Yes, she learned; but her fear of going into the cold Florida water in January kept her paralyzed in fear. Of course, there were high tides of frustration and anger as well. She'd spent so much time and money: swimming lessons, snorkeling gear, the very cost of the trip! This was an awful end to a desire that would not be met. And of course she'd always wonder if she'd made a mistake not taking advantage.

But as she looked at the water before her in Florida, she knew that she had to at least put her face in the water while she lay on the floating platform. She could do that, couldn't she? At least that.... Sadly, though, the group sponsoring the trip told her that the dolphins would not come near the platform. Those people participating needed to swim way out into deep water to swim with them. Eva was crushed.

But...what have we all learned about the heavens and desires that are placed into us? Miracles happen. So Eva, still hopeful but prepared for disappointment, put her face into the water.

(You'll get the chills here, folks.) Eva was overcome with emotion because there, eye to eye, was a dolphin – it seemed to look into her very soul and she, too, felt a connection that she can never forget. She says:

> The dolphin's courage to come near the platform helped me to summon my own courage to finally get into the water and have a transformational experience with them.

There is a message from the hearts of dolphins that cannot be denied when they know of one who should hear it. Eva had been chosen to spread their love through her artwork and stories. Another miracle spawned.

A Timely Rescue

T. Stokes, holistic palmist extraordinaire from the UK, remembers a story told to him by his father who was stationed on a warship during WWII. T.'s father remembers that dolphins would often swim behind the ships and play – we've all seen them jumping in and out of the water as though there were not a care in the world. Their joy is contagious. The men on the ship would not only enjoy the antics, but would throw leftover food scraps overboard to them.

On one unfortunate day, the ship fell under air attack. A man on deck watching the dolphins play was washed overboard by a huge plume of water that shot up over the side. The ship was in terrible danger and they could not stop to save the man—he was left behind to suffer the depths of the open sea as the ship moved away. T. says:

> The attack was like many at that time, spasmodic and nasty, but repairs were attempted at sea and the name of the man overboard was included with the names of the dead at a small prayer meeting held later by the ship's padre on deck.

An odd thing happened, however, that amazed all who witnessed it. The man who had fallen overboard appeared back on the ship with a remarkable tale. T.'s father related that a school of dolphins, so the man said, took care of him and helped him stay afloat. They took turns nudging him from beneath the water. The man held onto them as he went into and came out of consciousness. In fact, the sailor said that these dolphins actually talked with him and told him that they were taking him to the safety of land.

As these stories go, the ship's captain of course thought the man to be confused and not in his right mind, but T.'s father assured the captain, "We have just given him a clean bill of health, mentally and physically."

But telling the man that he was going to safety was not all these dolphins had to say—oh no. They went on to tell him that all the sea creatures felt that man had made quite a mess of things and that they were here to help save the planet.

Of course, this dialog landed our dolphin whisperer locked away from the others. He was, however, allowed a thirty-minute exercise period. At this time, when he went to the back of the ship, the dolphins would go mad with excitement! How glad they were to see him!

T.'s father never knew what to make of it (as I'm sure most felt that way at the time), but we know now that these guardians are indeed here to impact us, and saving one man from a sure drowning is one step closer to saving the next ... and the next... and the next.

A Not So Gentle Plea

From Mark Schaedel, lover of all Nature and honored student in Native American beliefs:

> These [dolphin] elders met me in spirit ...then personally greeted my wife and I on tour on the Indian River. In short, thousands of dolphins are being captured, then sent to places ... for money. They are separated from their families and their natural habitats. Not only that, the ones not suitable for show are slaughtered for food. This is the thanks we give for their kindness of countless times risking their lives to save us from shark attacks and guiding us to where fish are and sharing their loving higher consciousness with us. These waters are constantly being polluted by phosphates and mercury. ...Ironically, those who eat them get a high dose of mercury. I would encourage you to ... support those who seek to keep dolphins in their natural habitat. Keep them protectively away from people who would capture them to harm them. Please do not eat them nor encourage this industry [entertainment].

Autism and Savior Animals

"A kitten is in the animal world what a rosebud is in the garden."

~Robert Southey
English Poet

Another Animal Communicating Ghost Hunter

We were happy to be of healing help to another human being after the writing of Dinah's first animal book, detailing the how-to and learning about animal communications.

A ghost hunting friend of ours was the first in her family to read the book and to suspend any disbelief in the varied ideas related to animal communications. She says:

When I sat down to read your book Saturday afternoon I had high expectations for both an enjoyable read and some interesting stories. What it turned into was a life-changing experience – for my son. He is twelve, high-functioning autistic, and is, simply put, wired a little differently then most kids.

I knew by the time my son turned two that he was mildly psychic, but the signs were gone for many years until he hit puberty last summer and started seeing spirits. Not often, mostly around our house, and both human and animal. He loves animals and often relates to them better than he does to people, so I knew your book would be right up his alley. I read him the back cover and he was hooked (I got one chapter in before he confiscated it...)

His first goal was to meet his animal spirit guides – which he did. He varied the journey to make it his own, and talked to them several times, learning from each their names and what they were there to help him with. Being in mommy-mode, I prodded him to use the same method to meet his human spirit guide. He agreed – he would eventually – but I could tell he had little interest.

Tonight my son hit a bit of a rough patch and then said he felt sick – a common outcome that he's been struggling with. So I tried something. I told him to do the same thing you'd taught him, but not to go to the forest to meet Hy and Rough [his animal guides] – instead, to just to sit quietly on the beach and watch the water for a little while.

He came up with the idea of using some white noise on his iPod to help drown out the world so he could focus better, and off he went. Now I can't say which happened first – me asking his spirit guide to meet him there on the beach, or me knowing that his spirit guide was planning on doing just that. Five minutes later he opened his eyes and said, "Well that was unexpected!" I had one big knowing smile on my face.

He told me that when he got to the beach, he saw two chairs, so he sat down and looked out at the water as he had planned. Maybe thirty seconds later, he noticed someone sitting in the chair to his right. The being was transparent and glowing – a soft bluish white – and it was difficult to see his features. That's not a description he's used in describing any of the spirits he's ever encountered. The young man told him that he was Steve, his spirit guide, and explained what he was there to help him with in his life.

My son has always wanted to be able to communicate with animals and I have no doubt he will be doing just that soon enough. But more than anything I've always wanted for him to be grounded, to have a foundation to help him deal with the challenges ahead in his life, and to help him interpret any psychic experiences he may have. Little did I expect all that when I curled up with your book!

Well, of course, we were so happy to hear that this healing energy had been there for this young man, but that wasn't the last of it, according to our ghost hunting mom!

He has been bugging me to come on one of our ghost hunting investigations, but I doubted that the team would be too thrilled to have an ADHD 12 year-old along. I hadn't mentioned the animal communication thing to him lately – I wanted him to come into it on his own. In the meantime, I got this new case where a Golden Retriever was being tormented by a ghost and keeping everyone up half the night in the process. I mentioned the idea of an animal communicator to the client and she bit, so I had her send me a picture of the dog, Mac.

I was thinking maybe I'd bug you and then it hit me – duh – my son can do that now! (Well, maybe…) Of course when I told him I needed his help on a case he was ecstatic – ghost hunting AND talking to a dog – awesome! I gave him just a little insight and the picture of Mac and armed him with a few questions. He came back with some very interesting details. Nothing that would convince a skeptic, mind you, but interesting.

At one point, he asked Mac if his mom and dad were nice, and Mac said yes, they were, but they yelled "shut up" a lot. He was sure he'd gotten this wrong – he felt they were really nice people and wouldn't yell at a dog like that, but of course I had to ask them. The client got a real kick out of it – she apparently inherited a very noisy Conure and she was always yelling at it to "shut up." (When Mac was later told that he was not the one being rudely hushed, he simply said, "Oh.")

The client also has a second dog, whose name I hadn't been given, but I knew it was Mac's daughter. I only told my son that there was another dog in the house, but gave him no details. He wanted to do a test and see what he could find out on his own.

So he went into his meditation and came out a few minutes later all excited, telling me that it was a small dog, a bit like our terrier, but brown and gray in color, etc., etc. (Being autistic, he's excellent with details.) I think I broke his little heart when I told him it was actually another Golden Retriever. But it was a good learning moment...

"Honey, this may not be perfect, but most of the time it really works and you need to trust that. And we may not always interpret the things we see correctly ... Why don't we find a picture of a dog like you saw and we'll see if it means anything to them?"

So we scoured Google Images and found a Yorkie Poo that was close, although its muzzle was too short and it was a bit too small. I shared the picture and the story with the client - and said a prayer - and whew! She was amazed. She said she hadn't mentioned it, but along with the Conure, she'd inherited a schnauzer that looked a lot like that picture. Sadly they'd had to put it to sleep last fall.

She sent me a picture of ol' "Barky" and as soon as he saw it, he was all, "That's it!" and "That's the dog Mac showed me!" And what a deep sigh of relief came from me. I think we've created another animal communicating ghost hunter...

If you'd have told me any of this would happen six months ago, I would have thought you to be bonkers. My son, of course, would have told you he expected it all along.

Now we ask you, how can you not beam at a story like this?!

Mr. Scott, "Beam us up."

Another story of an animal coming to the rescue in an autism case is Mary's story of the kid.

The Kid

Mary has been blessed with being an Empath. For those who are not familiar with exactly what an Empath is, here is a definition to best describe. An empath is a person who has:

> ...the intellectual identification with or vicarious experiencing of the feelings, thoughts, or attitudes of another.

Now that you know what an Empathic person is, lets move onto this story about two of Mary's favorite people.

Since she co-hosts and supports paranormal and mind/body/spirit Internet radio shows, Mary participates in many chat rooms. Chat rooms can be very dangerous for an Empath. The whirlwinds of emotions, plus physical ailments can overpower in a heartbeat. But, once in awhile, a beautiful person(s) stands out, you start to chat, talk, and later become friends. Two in particular are... "The Kid" and his mom.

Around fifteen years ago, The Kid (his real nickname) lost his dad to a quick illness. For most families, the grieving process goes on with tears, depression, lots of memories, and expressions of feelings. But, when you are autistic, like The Kid, the grieving process is much more difficult. Your heart is breaking, but the ability to express these feelings becomes difficult. Times were very difficult for The Kid until a very special friend came to his aide. This friend, (not just any friend) was Mr. Scott. Mr. Scott was the family cat. The following is both Mom and an autistic child's tale of how Mr. Scott became the sounding board, great listener, and The Kid's best friend.

The Kid's mother tells us:

My son has always faced challenges. Autism has made every day a minefield of potential explosions. He doesn't know how to deal with his emotions, can't understand anyone else's feelings, and has no real clue as to how to relate to someone he does not know very well. He had to be taught empathy and sympathy, but he learned it well. His teacher, his hero, and the love of my life – his father – taught him.

His father and The Kid were attached at the hip. Where one went, the other was in his shadow. They liked the same things, did the same things, and understood one another better than I will ever understand either of them. My son was his Dad's kid and, though I occasionally suffered a pang of jealousy, I never regretted a moment.

When we got Mr. Scott (the Great Red Cat, as my husband called him), our son did not relate to him very well. He only watched him throughout his kittenhood.

As he grew, though, and it became obvious that he was an easy-going cat, my son began petting him, feeding him, and taking care of him. But Mr. Scott was *my* cat, and I was the one he sat with and seemed to want to be with. He would go to my son to be caressed once in a while, but they seemed to live separately.

My husband died in 1998. I worried about my son because he could not express his emotions about it. I could see that he was sad and I knew that he needed comfort, but he would not take it. He doesn't like physical touch and could not put into words (to me, anyway) what he was feeling. I didn't know what to do.

Then I noticed something odd.

I had placed my husband's urn in the living room. It is not a room we use very often and has no TV or radio. I began to notice that my son would sometimes go into the living room and sit on the love seat beside the bookcase holding the urn. He just sat… he didn't want to talk to me. He didn't want me to sit with him. So I would watch him and my heart would break all over again.

About a week or so after he started spending time in the room alone, I started to see Mr. Scott sitting beside him on the arm of the love seat. As I watched, Mr. Scott would lean against his arm or rub my son's cheek. He kept at it until he would start to pet him. Eventually, my son started talking to him. I don't know what he said. It was not my conversation. But he talked and talked, and Mr. Scott sat and listened. He never wandered away. He never made a sound. He never did anything but sit and listen. And my son became less sad. And then he got happier. Also, Mr. Scott was now my son's cat. He still loved me, but he stayed with him.

I don't know what they exchanged or how the cat knew how much he needed someone to talk to, but I know that I loved that amazing, beautiful cat. I miss him a lot, but I am blessed that he was here, and grateful that he knew what to do.

Hearing the story from a mother's perspective is very heartwarming. Most of us understand the feeling of helplessness and the feeling of gratefulness when someone (or some cat) comes to the rescue. Even more compelling is hearing the same story told through the eyes of the autistic child.

How "Scotty" Proved to Be A Loyal Companion – All His Life…

It was a beautiful spring morning during first grade, when morning game shows and nighttime comedy on TV were the rage. The school year had ended and Smoky had proven to be quite a nuisance. During that time, my dad decided to welcome a pet that was friendlier around people. He, my mom, and my grandma decided to take me to the pet store in E-Town to look for a loyal companion. By the time we reached the pet store, Mom found a sweet male red tabby looking for a home. When I walked by him, he just kept purring at me and reaching to me with his paws. Dad and I looked at other pets before we reached our decision.

When we brought the red tabby home, he had a little trouble adjusting to the new setting, but when he found me, he was all purrs and smiles. Immediately, Dad named him "Mr. Scott" after the character from *Star Trek*. He always had a penchant for naming his pets after *Star Trek* characters. I was getting to know Scotty, and slowly, but surely, he grew attached to me. Every time he meowed or purred, I petted him and he knew that I was his favorite guy. He also listened to me when I talked to him or called his name!

It was a dark day when I returned home from Louisville. Smoky, the gray tabby cat who was taken away from his mama too early, was dying near the washer. Immediately, Grandma was crying over the loss. Thankfully, Scotty was healthy, and he kept me company. He was in my lap, and while I was trying to get over life without Smoky, he affectionately raised his paws toward my hands.

I said to Dad, "I can't believe we lost Smoky!" Mom felt very upset over the unexpected demise of her respected gray tabby.

Grandma cried, "I don't think we'll be a one-cat household for long!" A couple of years have passed. She was still reeling over Smoky's death. However, Scotty formed a bond with her, too.

Things were getting rather unusual for Scotty back in 1993. After we moved to a new house, he was walking around the home, trying to get used to the new setting. He scratched at my bedroom door some nights when he felt lonesome. My dad picked on him when he felt sad, so he could be happy again. When I sat down in Mom's recliner, he would jump onto one of the padded arm rests and I would talk to him when I had no one else to talk to.

Scotty was a very playful cat, and he ate a whole lot of food when he was hungry. He used to sit by me when I was watching TV, working on the PC, and even playing games with my dad. Whenever I felt lonesome, he came by and brushed his furry body against my leg. He listened to me whenever I called him, and I oftentimes talked to him, too. He was the greatest pet I have ever had! I have one phrase to describe him: TRUE COMPANION.

The bond between him and myself really kicked into high gear in 1998, when my dad was stricken with a sinus infection. About every hour, Scotty was listening and walking right beside me at home. I talked to him like when I talked to Dad. Scotty purred and meowed back at me in a loving tone, making me feel less worried about my dad. I had to take a break from school during that time, but when my pals weren't there, I spent a lot of time with Scotty. When my dad suddenly died from bacterial meningococcal meningitis on March 1, Scotty and I cried uncontrollably. However, when the grief subsided, he and I continued to talk to each other. Scotty sure calmed me down in even the scariest of situations.

In the early 2000s, Scotty's health took a turn for the worse. I immediately grew worried about him when he stopped coming upstairs. Then, when I went to the utility room to wash my clothes, I saw Scotty's lifeless body.

I immediately yelled, "Mom! Scotty's not doing too good! He's in the utility room!"

Mom said, "Is he dying?"

I answered, "Yes, he is. He isn't dead yet, but it won't be long!"

After a few days, the sad news was confirmed. My dear cat Scotty had passed away. I felt so terrible and was so grief-stricken, that my life would never be the same. Life still goes on for me, and I am still nowhere near over the loss of Scotty. I miss him so much. I really do. In my sleep, I have dreams of that sweet cat lying beside me and purring. He is like a whole other person to me. Scotty, in his lifetime, made me feel so calm and so relaxed, that even Autism couldn't keep me from succeeding in life.

Our Speculations

So what are we to surmise of autism and animal caregivers? Were these gracious creatures put on this earth for this particular job in some cases? Are they medical rescuers or psychiatrists in the making? Many who have been involved with such cases will say that, yes, their special animal connections made such an impact that not only did the family who was immersed in the situation bravely move forward, but so did everyone hearing the love relayed in the story.

For our thinking, animals help autistic people. Period.

Are You Calling ME a Cat Lady?

"I kind of imagine myself at eighty, a cat lady."

~Juliette Lewis
American Actress

Who decides when a lady becomes a cat lady?

Why are only women portrayed as cat lovers?

How did such negativity come to be associated with people who have multiple cats?

Who was the first cat lady?

Who coined the phrase, cat lady? Was is some distressed man unable to reach his mistress in bed one night due to the slumbering purring creatures between he and his lady fair? Was it the lowly dog lover who had no compassion for a woman who talked baby talk to small kittens in a basket as his own Dobermans drooled into the basket with more interest in them than him?

From *The Seattle Times*, January 19, 2010:

> **Egypt announces find of ancient cat goddess temple**
> Archaeologists have unearthed a 2,000-year-old temple that may have been dedicated to the ancient Egyptian cat goddess, Bastet, the Supreme Council of Antiquities said Tuesday. The ruins of the Ptolemaic-era temple were discovered by Egyptian archaeologists in the heart of the Mediterranean port city of Alexandria, founded by Alexander the Great in the 4th century B.C.

An exciting find in Cairo for archaeologists to be sure, this was thought to be the temple of Queen Berenice (the wife of King Ptolemy III). He ruled Egypt in the 3rd century B.C. For our interests in cat history, the Egyptian archaeologist, Mohammed Abdel-Maqsood, said that there were a large number of statues of Bastet found in the ruins. This might mean that the temple could be the first dedicated to the cat goddess to be discovered in Alexandria and would indicate also that the cat goddess was worshiped into the Ptolemaic period. Cats go way back.

This goddess Bastet…just who is she and what did her cats mean to her? Known as a protector goddess and even a defender of the pharaoh and the sun god, Ra (not to be confused with the Star Being group by the same name from *Law of One* fame), images show her as a "fierce lioness" or a "woman with the head of a lion." Later, in the first millennium B.C., cats were becoming popular as pets. It was then that Bastet took on this representation of the head of a cat and became renown. This was a very good thing for the cats of our distant past because they became sacred animals. (We believe they still think of themselves in this way today. Don't you?)

So while it may be true that the "cat lady" comes from royalty, the term doesn't seem to imply monarch status in today's society. Rather, it brings to mind crazy.

Animal Hoarding

We have to take a moment to say that we are not talking about hoarding here, or any real medical problem that plagues some individuals into keeping cats of such great numbers that they cannot be fully taken care of, but we want to touch on this topic for just a moment anyway. (You see, we do know that there will be those out there who will use this against you, if you try to protect the "cat lady" reference.)

We believe we can safely say, after reviewing the research, that in most cases, animal hoarding people are oftentimes just overwhelmed care givers—thinking they are doing a good thing, but other things are going on in their brains that require professional help. At one time, these people were called cat ladies (when hoarding cats), but now a name has been given to the practice - animal hoarding. Thousands of cases are uncovered every year that involve varied animals (cats and dogs and even rodents), and organizations are doing their best to help both the people and animals afflicted.

Still, as long as there is no cruelty involved, most people on the outside looking in do not see this as true animal abuse (in that physical abuse is unusual) - it's not a crime. People, shaking their heads, often think the crazy cat lady is endearing. They feel for her because we all know she loves those cats, and the research says that she usually does. (This, by the way, is how the town drunk was initially viewed, according to Animal Behaviorist Steven Zawistowski from the ASPCA, who notes that just as with alcohol, there are victims here as well that go beyond the hoarder.) Stereotyped as a middle-aged woman from a good educational background, often a professional, with an attachment disorder and a dissociative illness, there are many other factors as well that create a hoarder (like suffering a crippling loss) and can push someone into this kind of behavior.

Though we are not talking here about animal hoarding as part of our "cat lady" phenomena in this light treatment of the moniker, there is a certain responsibility that should not be taken lightly for this particular sighting. Should you see a person involved in the hoarding of any animal, you should contact your local shelter or authorities for assistance for that person and those animals.

NBC News Correspondent Jill Rappaport tells us that if you see poor conditions in a home that has at least twenty animals, this should be considered a time to make the call. Other things to look for:

- Obvious lack of veterinary care
- Malnourished animals
- Animals not neutered or spayed (lots of babies)
- And as mentioned, 20 or more animals

Back to Our Kind of Crazy...

You've heard someone called a Cat Lady...and you may have even used it (in jest or otherwise). But we don't believe it is entirely based in fact, for there must be men out there in the world who adore cats, though you don't hear about "cat men" now do you? (Well, except for the legend of the Cat Man in Phoenixville, Pennsylvania - but he didn't keep a lot of cats. His thing was cat *food*. See the book *Ghosts of Valley Forge and Phoenixville.*) We do believe that certain men love cats as much as women. (We happen to live with husbands who fit the bill in many ways. Dinah has only had one cat, but she does know that if she took a basket of kittens to him and waved them under his nose, saying that these small darlings had no home...well, they'd have a gaggle of cats. He would most graciously take them in "until we found homes for all of them." Ten years later those same cats would be eating ice cream off his spoon. Mary has four and a revolving door rescue that her husband happily - most times - tolerates quite well.) So the first myth to throw to the gutter would be that cat ladies are alone in the world. There are cat men as well. (Now don't be sending us email telling us cats don't come in gaggles or we'll be forced to tell you a story about geese. And if you haven't seen a cat flying when a bird cage falls...well, ya just had to be there...)

But for our purposes here, what we see, first, is that the term *cat lady* may be used inappropriately or even as negative slang to describe someone who is misunderstood by the speaker. (We do admit that it can be used as an endearing moniker as well, though, depending on who is using it.) The conversation often stems from the question: How many cats are too many cats? The answer to this gives the speaker the ammunition to either cat

call or not (cat call…hmmm, another interesting phrase that has men calling out to women with lewd thoughts...) and, to those we've researched, many are quite opinionated. This shocked us, as we really do believe that the *how many* question is strictly a personal choice (unless there are issues of hoarding where everyone involved is at risk). All kinds of things were mentioned as reasons for having, or not having, cats. For example:

- Consideration of the size of living space
- Financial consideration of feeding and care
- Medical attention and veterinarian costs
- Amount of cat hair on furniture
- Amount of time available to spend with each animal
- Ability of cats to get along with each other
- The owner's personal allergic reactions

All, of course, are true depending whose reading the story. What was interesting, though, was that those who had all the stipulations were the ones who were most likely to criticize others for rescuing or having more than one cat by using the above rationale as solid reasoning and being vocal about it. They were the cat lady name callers along with the people who either didn't like cats at all, or who merely tolerated them because of a loved one. Those doing the rescuing or having more than one cat out of love for them talked about the family-oriented reasons, having buddies for the other animals in the house, interacting with them, the love exchange, rescuing them from rash circumstances... These people often had five to seven cats without negative personality issues to deal with or thinking about any of the downsides. Now that is not to say that these people do or did not consider the above-mentioned rational things. They did and do. Those things are just the givens to them. The important things for them were the "love" issues.

We conducted a small survey consisting of about 50 random people—not a big number, but enough to get a picture. Let's take a look at the questions first.

1. Do you like cats?

We first wanted to know out of a random mix, how many people were likely to even like our fair kitty friends.

2. How many cats are too many cats?

Then we wanted to go to the meat of the issue. Would the opinion of people who had cats differ from those who did not? Did anyone really care?

3. Do you feel (if you have a cat) that you can communicate with it on a level beyond using your voice?

Of course, this book is partially about animal communications, so we wondered if the people involved with cats had communications with them beyond the norm.

4. (Optional and hooked to the prior question) How?

We made the essay question optional, because...well, we know that animal people like to be with animals and not writing essays! (At least that's what we thought, so we'll see how those results went!)

The Results

Out of a random sampling of 50 individuals, 32 responded. So for our figures, we will consider that 18 of the 50 individuals either don't care for cats, don't care for surveys, or just don't care. (Well, yes, it could be they don't like us, too. But hey, aren't we all supposed to be mind/body/spirit people here?)

Question 1. Do you like cats?

32 individuals indicated that they did indeed love or like cats, with one included who liked them but was unable to have one due to allergies, and one who liked them as long as they belonged to someone else.

Question 2. How many cats are too many cats?

There were many varied answers here.

Number answers include: 6+, 3, 2 (in house), 2 (per 1,000 square feet).

Other answers: It depends (with no indication of on what). Depends according to space and resources. Depends upon the number you can take care of.

Some quotes for this question include:

Is that a possibility? I guess I would have to say too many is when it gets to the point that you cannot give each of your cats the one-on-one time for connecting and playing that they deserve—or when your house is wall-to-wall litter boxes, whichever comes first!

Depends on your space and your resources. Cats should be well-fed, clean, provided medical care, and have a clean and adequate-sized space to roam. Too many is when you get to the point that you can't provide that for each cat. If you have 50 cats in your house, that's too many. If you have 600 cats on a cat rescue preserve that is not too many.

I had 5 at one time. That was a lot, but never too many.

Too many cats in one location are more cats than can be properly cared for by one person/family.

I have a friend who has 5 cats and a dog in her house and you cannot tell until you see a cat hiding behind the couch, getting ready to pounce, or the one who is hiding in the hall closet on the top shelf eager to pounce as well. (How do they get up there?!) However, I have been to homes where there are two or three cats and it smelled terrible and cats were on the counters and such—too many cats if you cannot take care of them and train them. Cats, like people, need structure and routine to a point.

Question 3. Do you feel (if you have a cat) that you can communicate with it on a level beyond using your voice?

18 said yes, 4 said no, 4 didn't know, 6 didn't answer.

Question 4. How?

Some of the responses were:

Our cats have been able to deduce when I am sad, upset, or anxious. Their own actions reflect this without me saying a word. I have had cats wanting to cuddle with me when I have been sad, avoid me entirely when I'm upset, and get fidgety themselves when I have been anxious. Sometimes when I'm stroking my cat, we'll look into each others eyes, and I'll get the feeling she's thinking the same thing I am which is, "What's going on in that brain of yours!" I know that my cats have communicated with me non-verbally, or rather, "non-meow-ably."

Sometime's I'll be petting our current cat, Bailey, and even while she's purring away, the pupils of her eyes will get large and she gets this faraway look, almost as though she's in a trance. It's then that I know mischief is afoot. Soon she'll grab for my arm/hand then run around and come back for more. Even on slightly overcast days, I knew when a thunder storm was approaching because our cat Fluffy would hide and not come out no matter how much coaxing or how delicious the proffered treats.

As a kitten, Fluffy started out as an office cat where I once worked. When business slowed and I had to find other work, she remained at the office with the business's owner. When he became ill and could no longer go to the office, I and others would go to the office to check up on Fluffy. She developed a terrible skin condition, later diagnosed as an autoimmune problem, whereby her luxurious hair (she was a Maine Coon cat) was falling out leaving weeping sores. Meanwhile, through the years, we had gone from having six cats at home at one time down to just one as they aged and died. At that time, our one remaining cat, Domino, was 16. I was concerned how bringing Fluffy home to live with us would affect Domino, but mostly I perceived that my husband wasn't interested in building up our cat repertory again. One day, he went with me to the office to "feed and water" Fluffy. When he saw her condition, he said we just had to take her in and get her healed. I was so happy and proud of his generous heart—we both knew this would mean huge vet bills. Fluffy only lived with us 18 months before passing on, but in those months, she showed such devotion to my husband, even over me whom she had known all her life. It was as though she knew he was the key to bringing her to live with us and she was thanking him for her second chance.

I could use a snap of the finger, or clap my hands to keep them off furniture or call them over to me.

My cat can sense my mood and will act accordingly.

No, they don't even acknowledge me when I talk to them with my voice.

My cat is just as happy to be pet, brushed, and looked at lovingly.

I snap my fingers and pat the couch and he jumps up.

Several times I've felt cats delivered messages and communicated with me with body language, looks, and purrs or meows. I'm sure my body language and looks have communicated with cats as well ... probably in some ways I am not aware of myself ... and might not want to be!

I think cats can tell by your facial expressions if you are dissatisfied with something they have done. Sometimes, I just shake my finger at them and they know they did something wrong. In the same token, a kiss on the top of their heads or a big hug when I pass by lets them know they are loved beyond being bad. I sometimes sing to them, but I better not give up my day job!!! They love me off key or not.

Now, it was our intention to have a bar graph once we'd collected the data, that was precise and gave some bottom-line statistics, but as you can see, the answers, though basic enough and along the same vein, all have more depth than simple *yes* and *no* responses. Are we to assume that this sampling of people would then correlate to the masses of animal lovers (and haters) of the world? We know that scientists often say *yes* to this kind of data if it meets a whole slew of parameters, but we don't feel so confident about that. Why? Well, the 18 that did not reply worry us. What we have is a non-probability market research survey (yes, statisticians have to make explaining it complicated—a dirty job, but someone has to do it) or maybe even a convenience survey for we didn't really exhaustively search out groups of people to conduct our assessment. And, would you believe, there are even complex mathematical equations involved with this?

The reason the 18 "no shows" bother us is because cat-loving people do not want to admit that there are 18 individuals out there in the world who do not care. There. It's been said. Yes, most cat lovers care about all animals, and even have large families of varied species, but the families who represent those 18 may be so shocked at this data, that they will respond after the fact thus making the entire process void. (Because in a cat lady story, all people love cats...)

We do what we must however, and below is the chart. We've chosen to show:

- The number of people who love or like cats
- The number of people who did not respond (Unk. = unknown)
- The number of people who feel they can communicate with cats
- The number of people who do not know whether further communication is possible or not

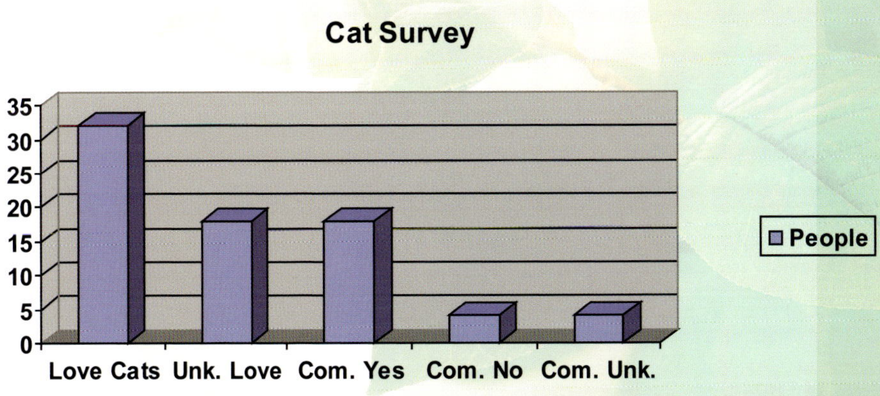

For our interpretation, the "Love/Like cats" category is a winner. (How did you know we'd go that way?) And can we use numbers to figure out just how many cats are out there in the wings? The American Pet Products Manufacturers Association (APPMA) is quoted as saying:

> According to the 375 page survey that tracks hundreds of pet ownership trends, Americans own approximately 73 million dogs, 90 million cats, 139 million freshwater fish, 9 million saltwater fish, 16 million birds, 18 million small animals and 11 million reptiles.

That's 90 million cats. Just here in the U.S. we were once told that there is no public place on earth that you can go that does not have cat dander. (We all carry it with us.) Scary, huh? So if we are ever invaded by aliens, they better not be allergic to cats if they expect to live here. (We guess we better say that we heard that dander statistic years ago and we don't remember where or from who and we're not sure it's true, but we liked the sound of it…so…)

The National Pet Owners Survey (NPOS) tells us that we are buying gifts (even in this time of financial crisis) of all kinds to please kitty. About 63 percent of cat owners are plunking down their dollars to make the felines of their families healthy and happy. We celebrate their birthdays, religious holidays, fun holidays (just think of the Halloween costumes…) and then there's that no-occasion-at-all holiday – the "just because." Do we need to go into the family bed, pet insurance, day care facilities, pet parties, travel and hotels extravaganzas, super meal programs, treats, toys, and their own special Internet pages?

Our point here, again, goes back to: What is a cat lady? Maybe there's more to it than a number and a gender. And if that's so…well, do you have a mirror handy?

Photo credit: Carol Starr

Ken Svehla of Ghostly Images advised Ghost investigator Carol Starr that downstairs in the basement at the Gettysburg Orphanage, he could hear a cat meowing. The basement is completely underground and there are no openings for a cat to get in. Also, they are smelling cat urine down there, too. Again all searches come up empty. Many people on tours also say that they hear a cat. Here see the steps that lead to the ghostly cat area.

Business as Usual

~ *Reported Ghost Cat Sightings in Public Places!*

You might be interested to know that in the following public places, the ghosts of cats have been seen or heard! So there may even be places where cats are haunting their cat ladies. Of course, these are just a few!

Battlefield Bed and Breakfast, Gettysburg, PA
Biltmore Estate, Ashville, NC
Capitol Building Basement, Washington, D.C.
Columbia College, St. Clair Hall, Columbia, MO
Courtyard by Marriott, Anchorage, AK
Doniphan Cemetery, Doniphan, MO
Elaine's Haunted Bed and Breakfast, Cape May, NJ
Fairport Harbor Lighthouse and Marine Museum, Fairport Harbor, OH
Gettysburg Orphanage, Gettysburg, PA
Hart Theatre, Ferndale, CA
Iron Island Museum, Buffalo, NY
Island Hotel, Cedar Key, FL
Little Crater Lake, Mount Hood National Forest, OR
Mars Railroad Station, Mars, PA
Molly Brown House, Denver, CO
North Court Cemetery, Ottumwa, IO
Old Coffee Pot Restaurant and Café, New Orleans, LA
Roby House, West Thompson, VT
Shanley Hotel, Napanoch, NY
Thayer's Historic Bed n' Breakfast, Annandale, MN
Thias House Bed and Breakfast, Washington, MO
Tucson Medical Center, Tucson, AZ
Wyoming Frontier Prison Museum, Rawlins, WY

Fluffikin, the Amazing Escaping Rabbit

"If you find a large tame rabbit in the garden, it belongs to little Gemma at number 7."

~A poster

Many of us, when thinking of healers, immediately go in our minds to the physical healings of medical issues. Our animals, however, think on a grander scale and healing comes to them naturally. Getting messages across that spur healing is not always as easy, though, since we humans…well, we don't always "get it" the first time. Or the second, or the third… Luckily, there are rabbits like Fluffikin to make the difference.

This next story comes from lecturer in Paraspiritual studies, T. Stokes … and we hope that you "get it!" T. relates his tale:

Joanne is a 30-year-old single mother, who has brought her daughter, Gemma, up on her own, after a short and abusive relationship with the father. This has left her with much emotional scar tissue, and, like many young women who have lost trust in men, Jo has developed a compensatory love for animals, and now takes in many of the local unwanted strays. Animals love unconditionally, and for many women, this is the only love they can trust. But a mix up at Social Services meant Jo's rent was not paid, so Jo, her daughter, and the dog, two cats, a pair of parakeets, grass snake, a tortoise, Fluffikins, the large rabbit, and Kevin and John the goldfish were evicted.

Now, no landlord would take Jo and her menagerie into their rented property, so Jo had to look far and wide, eventually finding a small cottage owned by a very old, understanding lady in a rather snooty village some way away who promised to accept all Jo's animals.

Many British villages have become little more than cold, suspicious, lonely, dormitory towns, as they no longer have shops, a post office, a pub, a church, public transport, or employment possibilities, yet one small shop remained, which was the hub of the village gossip. And when Jo, a single mother with a daughter and her animal collection arrived, no one was particularly pleased. The next few days, Jo felt very isolated when no one replied to her "good mornings" as she walked the dog.

But change was in the air!

Her daughter, Gemma, noticed that Fluffikins, the large rabbit, a real escape artist, had dug his way out of the garden pen. Jo had to run round and knock on the neighbor's doors and get them to check their gardens for the missing rabbit. Eventually, he was found and brought home. But each morning, he again had dug himself out and escaped – it became a regular thing for Jo to run round the village looking for Gemma's rabbit. One morning the Cub Scout team turned out before parade to do a quick search, finding her rabbit hiding under a neighbor's car.

Then something strange happened. Jo noticed things were changing when the grocery delivery from the village shop did not have her name and address on the box of groceries, but a drawing of a rabbit. And she saw a poster in a house window which said: "If you find a large tame rabbit in the garden, it belongs to little Gemma at number 7."

And while walking the dog, total strangers were beginning to say, "Hi Jo, how's the rabbit?"

Small children would knock on the door and ask to see the amazing escaping rabbit, and bring delicacies like a lettuce leaf or a juicy dandelion. A sheet with regular updates of gardens he had visited on his travels would appear in the village shop window, a picture of Gemma and the rabbit appeared in the church newsletter, the local paper carried the story about all the children who were now friends of both Gemma and Fluffikins. Jo was thrilled when a retired carpenter offered to make the rabbit a secure pen.

So the rabbit does not escape any more, and the gossip and excitement may have gone out of the village, but Joanne and Gemma have made so many new friends...and guess who has been invited as a star appearance at the summer village fête (provided a secure pen can be found)?

Animals are often underestimated in their healing abilities, but to heal a whole community must be a reason to celebrate!

Four of the Speedwell Wolves. Upper left wolf, Sarge, upper right Trinity, bottom left Chomp, bottom right Geronimo. *Photo Credit: Chuck Reiner*

The Wolf Sanctuary of Pennsylvania

"The wolf is neither man's competitor nor his enemy. He is a fellow creature with whom the earth must be shared."

~L. David Mech
International Wolf Expert

Pennsylvania is a treasure trove of treats for those who love animals, including those that are wild and untamed. Though it has been over 100 years since wild wolves have graced the lands of Pennsylvania, in Lancaster County there is a beautiful sanctuary that houses over forty wolves on twenty-two acres of beautiful wooded land. These wolves are the Speedwell Wolves.

The Darlington Family is responsible for this wonderful educational refuge that has created a habitat for over twenty years, allowing wolf packs (as closely as possible) to maintain similar situations as in the wild, while at the same time offering the public the opportunity to learn and understand what the realities of the wild really are.

Who are the Wolves?

Let us introduce to you some of the Speedwell Wolves:

Billy, Galahad, Chinook, Blazer, Charbanae, Lakota, Willow, Lucky, Hudson, Trinity, Thor, Frodo, Scout, Sarge, Winston, Chomp, Lincoln, Levi, Lucas, Lenape, Smokey, Friday, Solo, Dalilah, Dusty, Swazie, Cinderella, Glacier, Chipper, Merlin, Keisha, Tioga, Casper, Kojac, Turbo, Charlie, Destiny, Lazerath, Hope, Aries, Geronimo.

Dinah tells her story:

Tuesday June 29, 2010, 10 am Tour

I was excited about visiting the Speedwell Wolves and also about my daughter Angela's first outside-in-the-world excursion as an animal communicator. She, of course, was feeling that same mental freakiness that I'd felt early on – that everyone was going to know that she was crazy and that she was not really hearing or feeling what she was hearing or feeling. I still was plagued by this on many levels but just shrugged it off pretty routinely, now knowing that you just can't make these things up.

We started out right on time for the hour-long drive with plenty of time for any stops or issues and the GPS system told us that things were right on track. Until we hit the road closed sign – in the middle of farmland with no civilization in sight. There was time built in, but maybe not *that* much. Still, we swung around and took a new road (something I'd never have done in the middle of nowhere without a GPS or a companion) and were only a couple minutes late. By luck, we'd made it in time – well, the tour was already in full swing and there was just enough time to sign in, pay the entry fee, and hand over a business card introducing this new book we were writing before joining the tour in progress.

New and practicing communicator Angela got an immediate strange feeling upon moving into the small crowd of about twenty people looking through the fences at several wolves, who were more interested in the treats being thrown over the fence by the guide than any of the people watching them so intently.

She whispered to me, "That one right there…Hudson. He doesn't like the guide at all."

"What does he say?" I'd asked.

"I don't hear him talking," she said. "I don't hear any of them talking. I'm not getting it that way. It's something I feel."

Ah, I think. Another Mary. She's feeling. I never feel, so I know this is going to be a clash between my daughter and me because my questions will frustrate her for sure.

"Well, what do you feel then?"

"He doesn't like the guide, I said. Not at all, not one bit."

It was then that the guide moved nearer to the wolf Hudson and the animal growled.

The guide spoke right away. "You see?" he said. "Hudson doesn't like me."

Hmmm. So much for being late for the tour. A bit later I explained to the guide that we were late and asked why Hudson disliked him. He told me that another of the wolves, Murphy, had become ill. He'd been walking into walls and acting very strangely. He'd had to go into the pack's area and take him out by cornering him. As it turned out, Murphy had gone totally blind. At any rate, Hudson, never forgot it, or the guide, and has held the grudge. Here was a communication between the guide and wolf that needed little description – it was evident. But it was Angela's first out-of-the-box (so-to-speak) try.

Further along, we were introduced to Glacier, whose story began in the world of humans as a young woman tried her best to domesticate him as a pet. This did not go well – and I suppose this is as good a place as any to mention that it is not a good idea to try to domesticate a wild animal. The whole wild part of the description has a reason. You cannot, for example, *make* wolves do anything. They choose what they will do and they do things according to pack mentality. Our human alpha lady trying to have a wolf pet was pack leader. When a human male was brought into her home, Glacier was not about to relinquish his position. Things went violently downhill from there and Glacier ended up in the Sanctuary – which really was a good place for him after all.

I was very interested in the tour guide's stories because I have a little author inside my head that continually writes as I listen to everything. Angela, however, was more interested in just looking at the wolves and she had the added responsibility of corralling my 6-year-old granddaughter, who was wondering why this "zoo" was not allowing her to go inside with the wolves to pet them. (Never a good idea.)

So Angela comes up to me just as the Glacier story is finishing up and she says, "That wolf is saying over and over – not to anyone or to itself…I can't explain what I mean – but I just keep getting: *Pet me. Pet me. Pet me. Pet me."*

The guide had just finished telling us a few minutes before as Angela had been chasing my granddaughter back from the parking area that Glacier, because he had been used to having human affection, would have no one in his area to work without first giving attention to him. Otherwise, there would be consequences to pay. The wolves did not bite, but they would take small bits of skin between their teeth and pinch – this hurt! So with Glacier, everyone pet him first!

One of the things that we learned on this day was how little the wolves wanted to interact with humans. They really had little need or desire to have anything at all to do with us. Story after story was told about how they interacted among themselves and relied on each other. I found this to be true as well for my own communicating with them. Not having a strong empathy side to my own ability, I wasn't sure if I had gotten through to any of them or not. The night before I'd tried to touch base with them, as I always do prior to meeting any animal group like this, and did not seem to touch anyone at all. I'd met another communicator in Pennsylvania who'd become very connected to a wolf pack in the Midwest, but she was known as a strong empath, psychic, and also had a powerful spirit guide who provided connections for her. Angela asked me several times if I was getting anything because she (being new) did not want to say out loud what she felt for fear of being wrong. (How well I knew that feeling.)

I decided I'd give it another shot, though I was not thinking that I would be fruitful in the attempt. It was Chipper that I zeroed in on. This wolf was lying facing away from me, curled in a ball with his back against the fence. I used my method to meditate down and pictured Chipper in my mind. I then, with camera in hand said,

> "My name is Dinah and I am an animal communicator. I would like to take a photo of you and then tell your story to the world in a book. I would let them know the plight of wolves and how important you and all wolves are. Please may I take your picture?"

Chipper, letting me know in no uncertain terms:
You snooze, you lose...

Immediately, Chipper turned his head back around over his shoulder and looked me right in the eyes. He held the pose for about four seconds. I stood there with my mouth and eyes open wide in surprise that he'd responded, and my digital camera hanging by the strap and not even turned on. He then turned away. So I started asking again and pleading... I told him I'd be ready with the camera this time. He wouldn't turn again. So I just took a photo of his back. And I'm still telling the story and now you know that it's best to be ready and never to judge what you can and can't do; because if you *can*...well, you see how that goes.

I do think that these wonderful creatures warrant another trip and some more study. But if you are unable to visit, why not consider adopting one from afar? They have a wonderful Adopt a Wolf Program. The Sanctuary currently provides food, shelter, and veterinary care for over forty Wolves with no government or corporate assistance. So please, if you are looking for your "heart charity," this one warrants a look.

Post Script:

It was months later when my daughter called me from half way around the country. "I saw Glacier last night," Angela began.

It took me a moment to move the files in my mind around before I pulled the image of the wolf to my mind's eye. Angela had surprised me since we'd not spoken of the trip since she'd left months earlier.

"He's missing the girl who had him. He's thinking about her. That's strange isn't it?" she said.

I was thinking the same thing, for we'd learned that these animals were handicapped when domesticated and were only really happy in the wild. I said as much.

"Yes, but it doesn't mean he can't love or feel something for someone or something else, does it? How can we say that we *know* that?" she asked.

I had no answer. I suppose "connections" can happen for any species and we'd be fools to think that we have the copyright to that book. We both sent Glacier happy thoughts and hoped he'd let us in just for that brief moment of solace.

Sadly, Glacier has passed on due to old age, but he remains in our hearts.

Visitor Information: 465 Speedwell Forge Road
Lititz, Pennsylvania 17543
wolfsanctuary@gmail.com
www.wolfsancpa.com/index.html

Speedwell Wolf Billy. *Photo Credit: Chuck Reiner*

The Little Ringer

A close friend had a wonderful story to share with us that always made us laugh. It was her communications with her dog, Lindsay. At every chance, I'd say to whoever would listen: "Tell the story about the bells!" So, here it is, and it is a story that not only instructs us that you can teach an old dog new communications tricks (if he wants to learn), but that some dogs can be saved by the bell*!*

I'll start by explaining that this dog was scary smart. One look in Lindsay's eyes and you knew who the superior being was – and it wasn't you! Not the kind of smarts that would win obedience trials or a place on the telly, however. Told to sit, he would consider...*why should I? Can't you see the ground is damp, you fool?* No, his smarts were used exclusively to manipulate all around him to see to his personal needs and comforts.

Having spent some of his early years with Nips, a Jack Russell, he soon learned that a frantic and well placed yap or two would propel old Nips out of a sound sleep, leaping to see what the fuss was about – so that Lindsay could then have a lay on that nice warm bed he'd been eying.... This never got old and Nips never caught on until he grew too deaf to hear the alarm. Lindsay was also fond of releasing Nips from captivity with a well placed paw on the screen door lever. No frantic hopeful pawing here, just stand on the hind legs and flip, it's done! I had to replace the lever with a knob.

I also made the mistake once of knitting him a colorful checkered sweater with *loooong* sleeves in the front to cover those impossibly *loooong* Italian Greyhound legs. Well, it looked so funny on him that it caused uproarious laughter. Now most dogs get excited and happy when their humans laugh, but he positively GLARED until I took the thing off him. He didn't speak to me for three days after that. My usual punishment for a transgression.

A Bell is Worth a Thousand Words

When Lindsay was fourteen, I brought a new puppy into the house. The boy was slowing up a bit by then and growing quite deaf himself – he could just hear a loud clap or shout if you were behind him. He had also decided that, due to his advanced age, he wasn't required to follow proper dog bathroom etiquette. He pretty much dropped his little bombs wherever he happened to be, usually the kitchen. Luckily his diet of dry kibble made them resemble little acorns or sometimes walnuts.

I had heard somewhere that if you put a bell on the door so that it jingled whenever the new puppy went out, it was possible that the puppy would associate the jingle with going out and eventually learn to notify you. Since the puppy was about the size of a hamster, I hung the bell from the doorknob so that it just cleared the floor, and out we went several times a day, *jingle jingle*. People in and out, *jingle jingle*, you get the idea.

One day, I'm in the living room and I hear *jingle jingle*. What! Has the new puppy learned? But…the puppy is right by my feet. Well whadaya know, maybe the old one is going to let me know now! I let him out for a bit, in again, have a sit and…*jingle jingle*! Up again, and there's a walnut on the floor! Lindsay is staring at me, he paces over to his bowl and extends a paw and – he's out of water!

Well the old one never stopped dropping his bombs on the floor, but he DID use the bell frequently to summon me to: fill the water bowl (ring and point), fill the food bowl, cover him with a blanket (he would stalk back to his bed, lie down and stare at you for that one), let him out on the sun porch to lie in the sun (ring and tap on the door). Not a demure *tinklejingle* either, but a clashing *CLANG CLANG CLANG!!!* A dozen times a day already, he had turned me into his own personal dog valet!

And STILL the walnuts on the floor! When I was at work he had my neighbor, who looked in on him sometimes, answering to the bell! And all the time the look that said, humans are SOOO hard to train!

And so for three years I remained Lindsay's personal dogsbody, to the days when his advancing age made him profoundly deaf and quite tottery. He still managed to ring that bell!

Lindsay, all decked out in formal wear for a wedding. (Maybe he should have a bell tie!)

The little ringer has rung his last but I'm sure he is summoning SOMETHING in the afterlife. And after all this, the new puppy seemed to have been born housebroken, never went in the house or needed a bell to tell me, just goes whenever I let him out and is never a bother. Go figure!

Communications, of course, is in the method of those communicating – whether it be one to one or one to many. Here we had an internal family communication that worked to a "T" – at least for Lindsay!

Sylvester and Tweety?

"Always the cat remains a little beyond the limits we try to set for him in our blind folly."

~Andre Norton
American Writer

The next story is a hall of famer that comes from my own animal communication background. It was way back in the beginning, prior to my absolute certainty that I could hear animals and wasn't just a person with a strange mental disease. This is a story I call: "Sylvester and Tweety."

It was a stormy day - well, it should've been, but who knew that when it all began? Let me go back a moment to fill you in on just what my warm, quiet, cozy household was like . . . normally. My husband and I, both in our wish-we-could-retire mode, had two sweet dogs - Tessa was a 14-year-old, 16-pound peek-a-poo, and Rufus was a 4- to-6-year-old (we're only guessing) 29-pound shelter rescue dog of mixed descent who sometimes resembled Jabba the Hut and definitely lived for the food. Adding to this small family of four were three chattering parakeets: Peanut, Butter, and Jelly who lived in a large cage in front of our patio door. Also in our zoo, was a small ten-gallon fish aquarium stocked with six interesting community fish. Even with all these pets, things were nice and quiet. We were an adjusted family.

But as you know, things never stay the same. Change came like a slap in the face on that dark and stormy day, in the form of my (at the time) 22 year-old daughter, her then 8-month-old baby - and her three cats: Zorro (the baby-whining attacker), Bacon (the sweet chatty girl), and PeeWee (the holy terror). This was an emergency situation - the kind you don't turn away. Stormy. So our family changed. It was now: Me, my husband, my grown daughter, small granddaughter, 2 small dogs, 3 parakeets, 6 fish, and 3 cats in a two-bedroom condo with a loft.

Crazy as things became, we all settled into a comfortable... well, tolerant...home life with only a few animal incidents a day to deal with. (At the top of the list was dear Rufus who was determined to eat *his* food *and* the cats' food.) Everyone's personality was taken into consideration under these difficult times. Tessa, my oldest and most fragile dog, took to slinking around the house in fear of cats jumping out from hiding places to scare her. (The table cloth on the dining room table had to go. Neither dog could walk near it without being swatted by a cat hiding in its folds.) Tessa's tail had remained between her legs in fear since the cats came. Rufus, a young would-be hero if we'd let him be, chased the cats every chance he got and was then the recipient of batting about the face given by the three cats when they finally turned to fight. He then would come whimpering back to me in a big hurry as if to say: *Did you see what they did to me?*

The cats, too, had distinct personalities. Zorro was elusive during the day, but a crier at night - all night - and he sounded very much like a tortured baby might sound crying. (It was eerie.) He also spent much time in front of the fish tank trying to figure out how to get to the fish. Bacon, the little girl, had a sweet disposition and was never in too much trouble. She didn't bother the dogs, nor did they bother her. Even Tessa stayed put when she walked by. She talked to everyone consistently - had lots to say. PeeWee was a fearless monster - brazen, bold, in your face, and generally a lovable pain in the butt. I'm sure he must have been the cat in the photograph: Hang In There. This one struts his stuff to anyone who will watch. And if you won't watch, he'll get in your face, climb on your shoulders, walk on your keyboard, and just be in the way till you notice him.

So as you see, we were getting on well - except for one problem. The parakeets. The cats really loved the parakeets - and I don't think in a good way. We could tell immediately when we found all three cats sitting on top of the cage peering in at the three frightened and frantic birds. We moved the cage to the outside patio porch for the summer, away from the cats, while we were thinking how to solve the problem.

One afternoon, worried about the excessive outside heat, my husband - filled with husbandly knowledge as husbands often are - decided that the best bet would be to bring the birds in and set the cage on top of two wood video holders we have that are roughly six feet high. With cage perched on top, we would barely be able to reach them. If we couldn't reach them, either could the cats was his thinking. They'd be safe.

I remember frowning. No, this would not work. A cat could (would) jump up to the cage. (I remember hearing Bacon in my mind saying, *I can jump that…*) A mild, heated discussion ensued that I allowed my husband to win. The birds came into the air conditioning and to the high top of the wooden furniture.

Mesmerized by this wonderful development, the cats sat on the end table near the cage like stone lions with only moving tails to show life. "See," my husband said. "Just like the fish. They just watch."

I said nothing - could he not know that the cats were devising a plan to get into the fish tank while they watched? I'd shooed Zorro many times from trying to push on the tank with his paws. I could still hear the voice in my head as he ran scenarios through his little head - I'd thought at that time that I was crazy. Still, I sat quietly. Maybe my husband was right. The cats were orderly enough these days. PeeWee finally jumped down to the floor to sunbathe in a beam of light. Zorro returned to his favorite daytime sleeping spot under our bed - he'd found a way to open the bedroom door to get in. Only Bacon, sweet Bakie, as we called her, still sat admiring the fine chirping birds.

My daughter and the baby were out that day, so my husband and I relaxed, engrossed by the television, eating a paper-plate lunch from the coffee table. It was then - when we least expected it - that Bacon made her uncharacteristic decision. She leaped straight up in the air from the end table - apparently thinking that this would land her on top the cage. Not a good decision for her. Instead, she fell short and hit the cage full force on one side and knocked it into the air.

Things went to slow motion: Bacon turning in the air as she followed the cage out and down to crash on the floor. The cage nearly landed on the sleeping PeeWee, who lunged out of the way at the last moment. Both dogs barking, me screaming, my husband cursing.

The cage door busted wide open when it hit the floor. Jelly flew out and upward to the loft office, flying into railings and bouncing off potted plants. (Thank goodness he'd missed the ceiling fan which I'd rapidly turned off.) PeeWee, not to be left out of any crisis, darted up the loft stairs in a flash of fur, saying in his mind to my mind: *Not to worry! I'll get him!* (I actually heard that in my head.) Rufus ran after PeeWee nearly tripping me as I sped after the cat to keep him from having a bird lunch!

Bacon, scared by the commotion and not one to take stage center, skedaddled into the bedroom suspecting that there'd be a price to pay! My husband managed to gather the cage pieces and contain the other two birds – flying inside the cage battering the sides and screeching apparent obscenities. Tessa was howling by now as if the moon was present, I was still screaming (at PeeWee), my husband was still cursing, but more softly.

Once I got a hold of PeeWee, I locked him in the bedroom along with Bacon and Zorro – a master feat. He was not a happy camper about that. There were birds to be gathered! The dogs calmed as we put the cage back into working order. Then we had to drag it up to the loft so Jelly would see Peanut and Butter, and then go back inside. Any communicating with Jelly had long passed – either I was not getting through or he did not trust me. He did not want to go back into the flying cage of horrors. This took more than an hour to accomplish and even then it was more of a backing-a-bird-into-a-corner kind of ordeal – never easy.

Well, September came and the bird cage was back on the porch for just the time being (winter would eventually come), problem unsolved. But for now, though the cats look longingly through the patio doors at the three birds on their perches in the cage, we don't think they can open the sliding door. Although I wonder. . .

Brings to mind Sylvester and Tweety – tripled.

Epilog:

Not long after, my daughter moved to her own apartment and the cat-wanting-bird problem was solved. But once the cats were finally gone, I heard much skittering about in the cage and a unanimous "Finally!" from the birds. (I didn't believe that I'd really heard it, but regardless, those sentiments were mine as well!)

Tip:

If you are communicating with the animals, it helps not to be in crisis mode.

Highland Orchards Farm and Market's Very Own Dolly Llama

"Someone told me it's all happening at the zoo."

~Paul Simon

Singer and Song Writer

Highland Orchards, a working farm since 1941, just might be one of those best-kept secret places that people whisper about – this one in the hills of Chester County, Pennsylvania. I always tend to stay away from places that tout as a feature: Pick your own vegetables. (That sounds too much like work to me, though everyone else I know loves the feeling of getting out into nature and grappling with their food supplies.) At the Highlands, you can pick your own or find foods pre-picked for you. That alone makes the trip worthwhile, but there's plenty more at this quaint farm market, from the fresh produce to the decadent desserts their bakery provides. I mustn't forget that their "Best of Philly" Apple Cider Donuts are to die for. And yes, yes, there's the wonderful activities that make all feel like a child again: the hayrides and playgrounds and gift shop goodies.

But not to be forgotten is Highland's charming little zoo and the one, bigger than life, llama who is Queen for the Day, everyday. A beauty, she holds court over goats and fowl and other fair creatures as little people reach through the fences with corn feed to offer. Dolly was born in June 2002 and has been at the farm since September 2003.

Dinah's first trip there was not in search of a story for this book, but rather as an outing with her daughter and granddaughter. Anyplace that combined corn on the cob, pie, playground, and animals was a hit and she was expecting to have a relaxing afternoon. She relates:

Once there, it was hard to take our eyes away from the grand creature staring out from the fenced pen. This was the Dolly Llama and there was immediate communication between her and my daughter. Dolly was aware of Angela and aggravated that she was trying to talk with her. She glared at Angela, and my daughter received the thought that the lovely black llama did not want to get up from her seated position near the shade trees to come closer to the fence, did not want the corn feed she offered, and did not want to strike up a conversation. She was "in a mood." She turned her head away as if to dismiss Angela.

As for me, Dolly would have nothing to do with me. I wasn't picking up a peep, so I busied myself with feeding goats along with my granddaughter – after all, we were there for an outing. A bit later, I noticed that Dolly had come close to the fence Angela was talking out loud to her. I joined them.

"I asked her to get up, since she was sitting, to come over to see me, and she finally did," said Angela.

I moved closer to Dolly and in animal communication mode, said to her, "May I please take your picture?" She posed beautifully for me and I do believe it's one of the best photos I've ever taken – there was much emotion in it. Though I must say there was a sadness to it. It was my feeling that Dolly was both angry and sad that day and the photo seemed to support the theory.

Later, Mary looked at the photo I took and began to immediately get visions about Dolly, only these were past-life visions. There are many times that a past life or a future life will interfere or affect (negatively or positively) someone's current life. We usually think of this when associated with people, but the phenomena happens also with animals. In Dolly's case, Mary felt that in a past life she'd been a horse and had been with others of her kind during her life. This wondrous past life

made her sad (and angry) oftentimes in this life because she was now alone as far as others of her own kind. Some times were worse than others.

Further research into the species did show that llamas were social animals that preferred to be part of herds. This begs the question: What was Dolly sent here to learn in this life, living a life outside of a herd mentality? Or, was she just having a bad day?

Regardless, Dolly is one lovely lady that you should take some time to visit if ever in Pennsylvania. When you see how the children watch her in awe, you will see her impact on the world. Yes, she has an appropriate name, our Dolly Llama…

Visitor's Information:

Highland Orchards Farm & Market
1000 Marshallton-Thorndale Road
West Chester, PA 19380
Phone: 610-269-3494
http://highlandorchards.net

On the Wings of Butterflies

"Legend has it that whispering a wish to a butterfly, then releasing it to carry the wish, the heavens will make the wish come true."

~Journeyofhearts.org

As Dinah has walked the paranormal lines these past years, she has come across butterfly stories several times. The most prominent and heartwarming she wrote about in *Spooky York, Pennsylvania* and involved the spirit of a woman's brother who visited her in the form of a butterfly. Motivational speaker Wayne Dyer also had a compelling butterfly story of a deceased friend, Jack, that he often tells, coming back to visit as a butterfly – a comical rendition of a true connection of life and afterlife.

Dinah relates a story from her own experience:

The Strange Visit

Until this time of paranormal interest, I just saw these small flying creatures as beautiful integrals of nature, another vision of God's work. But as my third eye began to open, I found that butterflies seemed to have more of a "bent" towards the interactions between life and death - and more consistently of the changes in between. I began to hear more and more accounts of people having small, quickie experiences where butterflies slipped in and out of their realities like ghosts, here one moment, gone the next, but leaving behind a question that had both the elements of doubt and those of certainty. Still, nothing had happened to me that would cause me to change my initial feelings of this natural insect beyond myth and truth, nature and unnatural.

But then it did. Happen, that is. And as anyone who knows me will hear me say, "Once something happens to you, it doesn't matter what anyone else thinks or says; you're a believer."

A Historic Look

But first, before I tell you why *I'm* a believer, let's take a look at what people have thought in the past about these beautiful beings. In ancient Greece, the butterfly was called "Psyche," which means "soul." There have been lots of people who have seen a connection between butterflies and death or some faction of the afterlife. They've been thought to symbolize the soul, transformation (for the obvious caterpillar reasons), transition, time, lightness, celebration, and resurrection. The Irish have a saying that a butterfly is a soul of the dead and waiting to pass through purgatory. The Maya saw them as spirits of dead warriors in disguise. In Mexico, there is a town that celebrates the migration of the Monarch butterfly (during the Day of the Dead holiday) and sees them as the returning souls of their deceased. The Native Americans, too, believe that they are the souls of individuals who have passed on.

Perhaps my favorite take on this comes from the Aztecs, who believe that the happy dead return in the form of beautiful butterflies to visit relatives to assure them that all is well on the otherside.

A Protector Visits

And so my story begins. Sadly, my granddaughter, age 6 at the time of this experience, was without a father. He was killed in an unfortunate horrible incident through no fault of his own when little Ariel was just 8 months old. My daughter and I have felt his presence from time to time in many situations, but none like this.

We three were outside on my deck in the hot part of a sunny July day. Suddenly, a beautiful orange and black butterfly came to land on the porch swing next to Ariel, who was gently swinging back and forth. At first, none of us paid attention other than to comment on its beauty and to notice that it was very close. Then it "jumped" closer to Ariel. Then closer. We all watched with wide eyes as it then landed on her foot and sat still, seeming to look at her intensely. Child as she was, she became nervous and moved back with a startled motion and it flew away. We thought nothing about it, until moments later it returned to sit on top the grill near the swing facing Ariel. It jumped closer again. Ariel recoiled back not knowing what to do.

My daughter and I looked at each other and in unison, mouthed her deceased husband's name. We'd both gotten the message from "somewhere" at the same time. Was this animal communication? Spirit communication? Just a knowing? Or were we crazy? (All of the above, perhaps...) At any rate, the butterfly stayed with us for over two hours. It moved about the deck following the movements of my granddaughter in a most uncharacteristic way. We tried to make sense of it, but it was not acting like any insect we'd ever heard of – except those I'd written about in a ghost book. We were certain that this was the spirit of a deceased loved one, come back to visit.

We did not discuss these thoughts with the child. It was not the time for such a revelation. But my daughter and I knew that we'd seen her deceased spouse that day, and it was comforting to know that he was looking out for his daughter.

I know that now, whenever I see a butterfly, I look about me. Is this just a butterfly I wonder, or is someone near whom I've known in the past? If it goes about its business, I go about mine. But if it comes to me…well…

A Friend Takes Flight

Cece Wick, a Reiki Master, EMF Energy Healer, Ordained Interfaith Minister, and Certified Hospice Volunteer also has a compelling butterfly story to tell.

Cece tells us:

I was at the kitchen sink, finishing up the breakfast dishes when the phone rang.

"Hello?"

"Cece, it's Sara; I need you. I need for you to do for me what you did for your nephew."

"Sara? Where are you? Aren't you in Arizona? I'm not sure what you mean..."

Sara and I had been friends for over thirty years; first meeting as neighborhood moms when our kids were in preschool. Over the course of those years, we'd supported each other through divorce (mine) and widowhood (hers). Our friendship had gone through many phases, sometimes close and sometimes distant, either by circumstance or geography. We had much in common:

Both of our husbands were lawyers, we both were dedicated, busy homemakers and moms, both loved to knit, and we both loved to spend hours on our neighborhood's pristine private beach with our beautiful children. And when our children left us with empty nests, we both became successful outside the home – Sara as a trailblazer in our state legislature; the first woman in the country to serve as majority leader in both houses of a state legislature, and me, in the world of television station management. Over the years, we were in and out of each other's orbits, but the friendship remained, and we were always able to pick up where we'd left off, no matter how much time had passed. Now, with both of us in our later mid-life, the time we spent together was infrequent, as Sara had remarried a few years after the death of her first husband, and she and her new husband regularly fled the cold Vermont winters for the warmer Arizona desert.

"Cece, you know how you healed your nephew after his car accident? I need you to do that for me, too," she implored. "I'm in the hospital in Phoenix, in the ICU. I just had surgery and they found advanced ovarian cancer. Stage 4. They didn't even remove it all, so I'm going home to Vermont to get treatment there, and I want you to be there for me, to work on me."

I knew what she was asking: I'm an energy healer and an interfaith minister, and Sara knew the story of how I had helped my nephew, Kyle, recover from an auto accident that had damaged his brain stem. I had done energy healing on Kyle for several weeks after his doctors had advised his parents that such an injury carried only the remotest possibilities for recovery. With prayer and healing energy, Kyle eventually came out of his coma and today, although hampered by short-term memory problems, Kyle works, drives, and lives on his own. His doctor calls him "the miracle boy." And now Sara, a staunch, by-the-book Catholic, who for two decades had followed my journey on the mystical path, was open to asking for my help.

"Sara," I answered, "I'll happily do everything I can, but please know that it wasn't me: It was God working through me when Kyle was injured."

"Whatever it was, I want it," she replied.

Sara arrived back in Vermont a week later, and within days, I was at her lake-side home asking God to work through me for Sara's highest good. Not knowing whether or not it was Sara's soul's time to cross over, I didn't concentrate on healing her cancer; instead, I simply allowed the energy to flow, trusting that her soul knew just where to use the healing Light, and that I was simply the conduit.

In the comfort of her cozy guest bedroom, we shared healing sessions regularly for several weeks, both before and after her surgery that had removed all of the cancer that the doctors could find. Our time together was magical, and Sara remarked one day that, "This is probably the reason that we've known each other for so long, so we could do this together."

After a few months of energy healing, prayer, guided mediations, and chemotherapy, Sara was pronounced to be in remission: no cancer showed up on a CAT scan. We were elated, and so was her physician, a gynecological oncologist who was also an Egyptian by birth; a descendent of a culture that had practiced energy healing since the beginning of recorded time. Gratifyingly, he believed that the energy work had played an important role in Sara being cancer free, and encouraged me to continue to work with her.

So Sara and I continued to have healing sessions, not only as "insurance" to keep the cancer at bay, but to reverse the effects of chemotherapy. Sara, who'd always been beautiful, with a magnificent smile that lit up rooms, and who had a deep fondness for brightly colored clothes, was now a pale shadow of her vivacious self. Because she was physically worn down by the chemo, we now used energy medicine to remove the chemo's toxins from her body, in an effort to increase her vitality.

Unfortunately, one day several weeks later, as I placed my hands on Sara's abdomen during a healing session, I was jolted when I unexpectedly "saw" cancer. How I "see" these things I have no idea, but suddenly tiny black spots "appeared" to me in her liver. Not wanting to upset her, but very curious to know if these spots were indeed cancer, I casually asked Sara if she had a CAT scan scheduled anytime in the future. She told me that, yes, she had one scheduled for the very next week.

The CAT scan revealed the cancer spots in her liver, and she began another round of chemo. And thus began her ride on the merry-go-round that is the course of ovarian cancer. For several years, Sara was on that hellish carousel; her time divided between her homes in Vermont and Arizona, and between chemo treatments and remissions. During this time, our healing work together became sporadic, and, into the sixth year after her diagnosis, I didn't hear from Sara much, but knew that her remissions were lasting for shorter and shorter periods of time. I trusted that we weren't meant to work together; that her soul was working other things out, in other ways. Then one Friday, in the early summer, I received a call from a mutual friend who told me that Sara, after six and a half years of surviving ovarian cancer, was hospitalized and actively dying. I rushed to the hospital.

Sara lay in her bed, thin and weak. I hugged her and told her how sorry I was that this time had finally come.

"Don't be sorry, Cece," she said, "I'm not; I'm sick of all the chemo, the scans, the fear, and the worry. I had a hard time in Arizona this winter, and I guess I knew that I'd come home to Vermont to die this summer."

We did a healing session, a session that flooded her being with Light, filling her with peace, and helping her to prepare for her transition to the Other Side. It was a lovely time, and she asked me to return soon to do another. Her daughters were arriving for the weekend, and so we made plans for me to return on Monday.

But Monday apparently wasn't soon enough for Sara, because on Sunday afternoon, while at home waiting for my little 2-year old grandson to arrive for a play date with me, I was sitting in the living room when I heard some strange rattling, and then a loud crash, coming from the kitchen. I ran into the kitchen to see what had fallen – but nothing had "fallen"; a plastic toy with various parts was lying in pieces on the floor in front of my dishwasher, a full four feet from where it had been stored in the recesses of a toy basket, on the far side of a child's high chair. I was stupefied: That toy had not moved on its own; it had been picked up and hurled against my dishwasher. And in an instant I knew; I knew that Sara had reached out to get my attention. And although I'd received messages, and encountered other roaming souls in the past, this incident sent chills down my spine: Nothing quite so dramatic had ever happened before. I was rooted to the floor, and asked aloud, "Sara? Do you want me?" But I really didn't need an answer; I knew that Sara had done this to get my attention. Experience had taught me that during our dying process, our soul moves in and out of our bodies, busily traveling to the Other Side, and visiting friends and family on this side, just as Sara was now doing.

But I always liked to have verification when something out of the ordinary happened, so I picked up the phone and dialed a woman in New Jersey who'd helped me in the past with spirits who'd followed me home on several occasions, folks who'd made their transitions, but who'd not yet gone to the Light. As soon as Vivian answered, I described to her what had happened, and her first words were, "Wow! Someone really wants your attention!" And I thought, "Tell me something I don't know!" But instead, I told her that I had a friend who was dying, and asked her if she thought she'd crossed over, and had wanted to let me know. Vivian said she'd need a few minutes and would call me back.

As soon as we hung up, I dialed the phone in Sara's hospital room – and she answered. We spoke briefly, Sara telling me that her family was with her, and that she was trying to get into the already full Respite House, our local hospice center. We made plans for me to visit her later that afternoon. As soon as I hung up the phone, it rang, and it was Vivian.

"Cece, your friend hasn't passed yet, but she's getting closer, and that was her spirit who did that in your kitchen; she badly wants you to visit her again!" I thanked Vivian for confirming the conclusion that I had also reached, and said goodbye.

That afternoon, Sara and I had a magical healing session, during which I led her in the Buddhist Loving Kindness Meditation, a meditation designed to help create peace, to bless the ones we love, and to release the bonds of resentment, anger, and unforgiveness that we might hold towards others. It had been my experience that this meditation frees us of negativity, especially during our time of passing, and enables us to fully turn to the spiritually transformative and welcoming Light that is our real Home. Happily, Sara loved the meditation, and thanked me, telling me that it was "so beautiful" as she drifted off to sleep.

I left her room convinced that this was the reason that everything else had gone before; that thirty years of friendship, and the many healing sessions, were all simply laying the groundwork for this particular, profound, shared meditation experience. I was content; there was nothing more for me to do for, or with, my dear friend. Our work together was done.

On Monday, I commuted as usual to my job as a lone ranger Executive Director for small non-profit that was headquartered in Montpelier, our state capital. Thinking of Sara, I was melancholy as I passed the gold-domed statehouse, the same place where she had been so enormously successful.

Sara died the next morning, surrounded by her family, and enfolded in the peaceful and lovely surroundings of the Respite House, which she had entered barely twenty-four hours earlier. Her obituary ran the next day, and because of her stature, her passing was noted by the state's newspapers, and on the many news programs of Vermont's television and radio stations.

I was deeply saddened by her passing, but I'd known that this day was inevitable. I felt grateful that I'd played a role in her spiritual growth, and in her dying experience. But all the same, it was very hard for me to think of beautiful, vivacious Sara, with her dazzling smile and her vibrant clothes, as being gone forever. I looked forward to her funeral; a final goodbye to a dear friend.

Clearly, the Universe had other ideas, because on the day of her funeral, I was hit with a hideous stomach flu and was too sick to attend. I was horrified. I wanted to say a last goodbye, but I had to rely on my faith and trust that this, like everything else in life, was part of a larger plan, and that all was well and right on schedule. I was also comforted by the fact that our time together was complete, and that we'd had our own special goodbye, but I was still disappointed.

A week later, I was joined at work in Montpelier by my good friend, Chris, who often helped out in an assistant's role when things at work got too burdensome for one person. It was mid-morning, and we were sitting in the sunny office, on opposite sides of my desk, verifying membership lists, when I heard a gentle tapping on the window. I'd never heard a tapping on the window before; the office was three floors up, making it impossible for anyone to tap out a friendly "hello." Curious, I looked at the window, and was astonished to see a big, beautiful, bright yellow butterfly tapping its head against the pane and fluttering its wings. Aware that butterflies are universal symbols of rebirth after death, I was dumbfounded, but not speechless: "It's Sara!" I blurted.

Chris, with her back to the window, jerked her head up and asked "What?"

"Look! Look at the window!" I excitedly shouted, pointing at the window. "It's Sara! She came back to let me know that she's fine! And she came back to make one last trip to Montpelier, and take one last turn around the dome! I know it's her, Chris!

Look how bright she is!" We both stared for another minute, with the lemon yellow butterfly still fluttering her beautiful wings, stationary in her flight path, as if she was staying long enough for us to "get it," for us to recognize her. And then off she flew.

Neither of us said anything for a few seconds, and then I looked at Chris, "You know, it's too early in the summer for the yellow butterflies – they don't appear here until August.... It was Sara, I have no doubt about that." I smiled. "And she's wearing a bright color, as always!" We were stunned, smiling, and speechless, savoring the lovely and astonishing thought that beautiful Sara had found a way to come back one last time.

Plants and Animals and Humans: Oh My!

> *"A man is ethical only when life, as such, is sacred to him, that of plants and animals as that of his fellow men, and when he devotes himself helpfully to all life that is in need of help."*
>
> **~Albert Schweitzer**
> German Theologian

As you know, things don't always make sense and they don't always happen in sequence. So the day that Vicki walked into Dinah's living room to tell her about how plants have relationships with animals and humans, she was taken aback.

Dinah says: You see, I'd had an incident or two already that had made me suspicious of the whole plant world and its ability to have the same impact on the world as either an animal or a human, but I'd not expected it to pop up again.

But let me share with you first the event that prepared me for "Percival" and for the coming together of plants and animals and humans. It all began with a song.

The Nature of a Child's Song

My young granddaughter came to visit. Because my daughter lives so far from me, I'd not seen them for a very long time and I was so excited to interact with them again and to share some of the paths I'd been meandering down. Much to my surprise, my little granddaughter had a path of her own. Everyday she wanted to go to the grocery store to buy cut flowers. My daughter said this was a common theme, that she just loved them. While at my home, she spent a lot of time on my deck—not only playing with my dogs, but admiring the flowers that grew around it.

The flowers around our deck all had specific times for blooming. We had those that bloomed just in spring, some for later in the summer, some all season long… There was always something interesting and beautiful to look at around the deck. We have a lovely rose bush that blooms in the spring, but it was finished and horribly, the beetles had attacked the leaves, though we actively were trying to rid the plant of them.

One morning, as I was drinking my coffee on the deck, my granddaughter came out to again talk about the flowers at the grocery store. I mentioned that we had many flowers here around the deck and I asked whether she had spoken to any of them. She looked at me kind of strangely and in wonderment.

"Talk to them, Mom-Mom?"

I told her that yes, this was something that should be done quite routinely and that they grew better when people talked to them. They were, after all, alive. She was nodding her agreement and listening very intently.

"You might even like singing to them," I said. Now saying that surprised me. I'd never sung to a plant. Never thought to do that. This thought had been put there by the Light Beings I was sure. I recognized their nudge. Okay, I thought. Sing to the flowers. That had a nice ring to it.

My granddaughter nodded again and then went about her morning. Later that day, my daughter came up to me while I was working at the computer. "Mom," she said. "Look…" She was pointing out to the deck at my granddaughter. We both watched in silence as she stood on the deck in front of the rose

bush singing her heart out. We couldn't hear the song through the door, but she was using her arms, raising them in the air as though she were singing an aria. It was the most adorable thing I'd ever seen and my heart soared.

My daughter said, "Did you tell her to sing to the flowers?" I told her I had, and we both watched for about ten minutes. Then, her song completed, she went on playing with the dogs.

They went home that evening. It was rush, rush, rush to the airport - never a pleasant trip. It was dark when I'd returned home and very late. I fell into bed and woke late in the morning the next day. I got up, poured my coffee, and went out on the deck. Then I stopped short. My rose bush. The prior day it had been devoid of roses (for weeks). But now, there were buds all over it! I was shocked. It felt impossible (but I knew that nothing in nature was truly impossible - even this!)

Over the next few days, I had a second bloom on a bush that never has had a second bloom of roses since we'd been at this home. My granddaughter has a clear path. I called my daughter immediately and told her to get that child some gardening tools!

Such is the nature of a child's song.

Dead on the Ground, But Not for Long...

But the story does not end there. The universe has a way of sending a message home – no matter how many times it needs to be sent or how many ways. Until we "get it." I was of course in awe of my granddaughter's ability to grow roses with a song. How wonderful! But not everyone can do that, I thought. Not everyone, and especially not me. I did not have a green thumb and plants could certainly take me or leave me (but mostly leave me).

A few days later at work, I walked into a co-worker's office to see her plant sitting on the corner of her desk disheveled and pretty much dead. I would have never even noticed it in days "before the roses." The dirt was a mess in the pot and there was dirt on the floor. The plant itself was brown and wilted and yes, it was quite dead.

"What happened to your plant?" I asked.

"It was hanging by the window and fell the other day," she replied. "It's dead and I haven't gotten around to throwing it out."

When she said that, I got a certain ping in my head. I frowned. What the hell was that? Another ping.... Of course, I wasn't too surprised at pinging after the initial touch because once you've worked with channeling, well, you pretty much have an open mind for anything and especially pinging. The Light Beings were always giving me a sign here and there – mostly telling me when to keep my mouth shut during controversial discussions. They spoke to me now in a very simple statement. *Take the plant.*

I stood there for a moment, nearly shocked by the internal suggestion because anyone who knew me knew that I was very good at killing plants. They did not live for me for some reason. I don't know why, but there had not been a connection more than once or twice over the years between me and a plant. And those plants I had connected with were of the sturdiest of natures. *Take the plant,* I heard again. Alright already. I'll take the plant.

"Can I have it?" I asked her.

No pause at all. "Sure."

So off I went to my office with one dead Wandering Jew plant in tow. I sat it on the edge of my desk, put water in it and went back to my computer work. Several people came in during the day to ask me, "What's with the dead plant?" I'd always shrug and say, "I'm going to bring it back to life." I didn't know where that certainty had come from, but somehow, that's what I knew I was going to do.

Now this may be hard for you to believe, but I swear it's true. The next morning when I came to work, there were "stalks" standing straight up from the middle of the plant reaching towards the sun coming in through my office window. The plant was not only alive, but it was *standing-up* alive! I was flabbergasted. All day this day, people came in to see the wonder of my undead office plant. They were amazed. Of course there were those scientific types who immediately wanted to tell me all the science behind bringing a dead plant to life. (Like, it wasn't really dead.) But those of us who'd seen it up close and personal the day before, knew. There was more to this than met the eye.

On that early morning when I discovered the plant alive again, I sat there thinking about the roses and this Wandering Jew. *Hmmm*, I thought. If the plant is responding to me and my vibrations, could I interact with it on a deeper level? Why not try? I decided to try animal communications on it.

I went down into my visualization state and this time, instead of picturing my animal guide with an animal, I pictured the plant. At first it felt odd, but then I watched in my mind's eye the plant's stalks all turn to me as if I were the sun and it was getting its nourishment from me. The stalks were many and healthy. I asked the plant its name (not entirely sure what would happen). Then, as though there were a whispering voice in my head, I heard it.... "Percival." I was so excited that I popped right out of my visualization. I'd contacted a plant and it had given me a name! Percival! Plants were conscious!

I'm looking at Percival as I type and he is basking in sunlight reaching toward the sun coming in my window at work again. My husband repotted him once into a bigger pot because he's gotten so big; the first one was strangling him. People comment on his beauty all the time and don't find it strange that he has a name - many people name their plants. But no one really believes Percival's story now that it's been over a year since it began. It is too easy to "forget" that there is a formal connection. Much easier to believe it never happened. After all, there's no scientific proof....

It's Science If You Want It to Be

Before we look at the science that includes animals, humans, and plants as partners in the universal dance, let's look at research that has gone on regarding communicating with plants in general. It's long been said and proven that plants do grow better when talked to or when certain genres of music are played. This is not new. Research dating back to the mid 1800s has brought millions to the reality that plants might not be family members (just yet), but they responded in fashion to the same kinds of relaxation mechanisms that people did - positive talk and good music. Books were written and music released - a new business was born. Penn State University stated in their research in 2007:

> "...scientists at South Korea's National Institute of Agricultural Biotechnology proposed that two genes involved in a plant's response to light - known as rbcS and Ald - are turned on by music played at 70 decibels.... This is about the level of a normal conversation."

Going back to the day when Vicki began talking with me about the connection between animals and plants was the research that brought everything together. She, all along, had been one of Percival's main supporters and talked with him every morning. She wanted to know whether I knew that plants had a connection to animals. Well, I'd said…sorta kinda. I knew they had connection to us, but I couldn't really see my Spiderman or Louie talking to my roses or the dogs that frequented our office talking to Percival.

But Vicki had science on her side and I was all to eager to listen. It began with a doctor, researcher Cleve Backster, and the coined tag given to a particular communication going on behind the scenes but within (and without) our bodies called the Backster Effect. Backster, in 1966, though not really looking for the results he found, hooked a polygraph to a plant in an effort to see how long it would take water to get to the outer most leaves – an experiment on resistance. (I don't pretend to understand what he was trying to do and it doesn't matter for our purposes here.) What he got was something that appeared to him to be very similar to the data that would be displayed if someone were taking a polygraph. Then, according to an interview with Cleve Backster published in July 1997 and translated into Czech in October 1998:

> Then at thirteen minutes, fifty-five seconds chart time, the imagery entered my mind of burning the leaf I was testing. I didn't verbalize, I didn't touch the plant, I didn't touch the equipment. The only new thing that could have been a stimulus for the plant was the mental image. Yet the plant went wild. The pen [from the polygraph machine] jumped right off the top of the chart.

The plant was responding to the thought that a human had of destroying (by fire) one of its limbs. It could read minds. And adding to that, he found that if plants were exposed to humans or animals for even a couple hours, the plant would exhibit a connection to that other living being. Of course, things did not stop there. Then came more and more tests and responses from the plants on everything from advertising to genocide. Here is where the animals came in and it actually makes the uncomfortable connection between humans, plants, and the foods that humans eat – i.e., animals and vegetables (plants). The good doctor arranged for an experiment where he would drop shrimp into boiling water to see how the plant (presumably watching somehow) would react. Simple to say that once other living stimuli were removed from the plant's environment (thus keeping the plant from connecting with it), when the shrimp was dropped into the water, the plant screamed out on the polygraph.

The final point that Vicki and I discussed at length was another of Backster's findings dealing with the consumption of plants and animals in the human diet. Knowing now that plants reacted horribly to the death of any living thing (and this includes anything living with a cellular structure - which is everything), his experiments with dropping animals into water for consumption or cooking meat, steaming or boiling vegetables became a frightening thing. My gosh...all the screaming that must be going on at dinner preparation points around the world.

There was an interesting fact that came forward though from his work. If the entity about to be "killed" is warned prior to that death (drop in the water, etc.), it will "flat line" and feel nothing. Backster tells us:

> Eggs. I had a Doberman Pinscher for a while back in New York, and I used to feed him an egg a day. One day I had a plant hooked up to a large meter ordinarily used to display galvanic skin response. ... I could see on the meter any large change in reactivity. This particular time I was feeding the dog, and as I cracked the egg, the meter went crazy. I thought, *What's the connection between cracking an egg and the plant in the other room getting all whippy.* That started hundreds of hours of monitoring eggs. Fertilized or unfertilized, it didn't matter; it was still a living cell, and plants perceive when that continuity is broken. Eggs, too, have the same defense mechanism. If you threaten them, their tracing will go flat on you. Then, if you wait about twenty minutes, they come back.

So the idea then would be to warn the food you are about to eat, whether it be vegetable or animal....Vicki advises, "I've read that the food we eat feels that it's a privilege to be eaten by us." She suggests, and I concur and have already started doing this (but it's super hard to remember after all my human conditioning), that we all think about the food we eat prior to preparation and warn it first! I certainly don't want my plants and steaks screaming at some cellular level! Yes, yes, it sounds odd. But there's research on this one, folks. And what does it hurt to add this one caring thing to your daily rituals?

Speculation Aplenty

In the meantime, and back to animals and plants, I'm fond of free thinking and linking thoughts that seem to make sense. If plants can hear thoughts, have reactions, and foster connections to the living things around them, and we already know that animals, too, have these capabilities, why then would we not think that our house plants are not connected emotionally to our pets? Why wouldn't zoo animals be connected to their environments and caretakers? Why wouldn't sea creatures bond to the water's organisms?

Therefore, we move into the thinking of environments and how hurting connections – even if accidentally or unintentionally – impact animals (and now plants) more than we can ever appreciate. I'm looking at the living around me in an entirely differently light now. My family has expanded exponentially.

Think About:

Some Environments That Hurt the Connection

- Any oil spill (Exxon Valdez)
- Wild fires (California, Texas, Midwest, etc.)
- Disappearing of the rainforest
- Pollution (land, water, air)
- Habitat destruction
- Species extinction
- Toxins and waste
- Overpopulation

Short Animal Stories

Let's Hear It for our Friends!

"Humans are amphibians – half spirit and half animal. As spirits they belong to the eternal world, but as animals they inhabit time."

~C. S. Lewis
British Author

The Animal Communicator

We'd like to begin our short story section with a couple communicating incidents that happened to Mary. Every animal communicator has a funny story or two to tell. She has many and thought she would share two with you.

Daffee Goes Home

Let's start with Daffodil, a.k.a. Daffee. A while back, I was at our local pet shop buying food for our aquarium fish. As any other animal lover, I always strolled the aisles to look at what I wanted to take home next, but knew I couldn't.

Observing the pretty little parakeets preening themselves, I heard a tiny child-like voice from behind me say, "Look at me, look at me." I turned around expecting to see a small child and there was no one. I shrugged my shoulders and thought perhaps it was just the music coming from the ceiling speakers.

I went back to looking at the frolicking parakeets hopping from one perch to another and again I heard, "Look at me, look at me." I turned around and when I did, a beautiful yellow cockatiel was staring at me. She was all alone. We made eye contact.

That day, back in 1999, was my first true experience, that I can recall so vividly as being able to communicate with animals. I also left the pet shop with more than just fish food that day.

Daffee has become a major part of our family.

An Apple a Day

On another occasion, Dinah and I attended a Halloween function at a B&B in Gettysburg. On Halloween night, we were going into a haunted barn for a costume party. Right before we entered the barn where the festivities were to take place, I heard a voice that sounded like, "Hey you…you! Come over here."

I looked at Dinah and asked if she heard that gruff voice. She didn't, but noticed this beautiful horse looking in our direction. Not wanting to be late for the party, we kept on walking and louder I heard again, "Hey you! Come over here!!!"

I looked at Dinah; surely she had to hear that. It was so loud. But she didn't. We looked over at the horse and it was staring right at me. So, I did what any animal communicator would do and went over to the horse.

Let me now introduce you to Furry Berry, who is a beautiful horse that resides at the Battlefield Bed and Breakfast in Gettysburg. It seems Furry Berry loves his home, but there is one thing he is not too happy about. We all know horses love apples. And what do you think Furry Berry looks at all day long and can't reach? That's right, apples. The driveway is lined with beautiful apple trees, full of those red beauties, but Furry Berry can't get to a one. So, understandably he is quite upset.

I told him, I would see what I could do. Later on in the evening I told his owner, Florence Tarbox, Innkeeper, that he needed some apples. I am very happy Florence didn't question how I received the information, because sometimes talking to animals is very hard to explain to others.

On a recent visit back to the B&B, I met up with Furry Berry. He is very happy these days. His plea has been answered and he gets that occasional treat.

Visitor Information:

Gettysburg Battlefield Bed and Breakfast
2264 Emmitsburg Road
Gettysburg, PA 17325
888-766-3897
www.gettysburgbattlefield.com

Photo Credit: Rick Gasparo

We love being able to communicate with animals. Even better when it is message from them that has an easy solution.

It should not be dismissed that animal communication's impact on the individual reflects a connection on a much larger scale. One animal to one human opens up a smorgasbord of connections that cannot be denied. That impact spreads with each new communication, each new animal, and every time a story is told. For you see, animal communications are around us all the time – sometimes like these two prior stories with real voices, sometimes by behavior and antics, and often by just a gentle knowing that contact has been made.

So listen closely...and we invite you to read the impacting stories that follow!

Little Crab vs. Big City

Artist extraordinaire Ted Enik lived in Chinatown for a couple of years and tells a wonderful story of life, of crab. He says:

I lived in New York's Chinatown for a couple of years during the '90s. Daily, one could easily find a wide variety of allegedly edible flora and fauna being hawked right downstairs on the street. Bucket of frogs? When in season.

One crisp, early fall morning, after walking my girlfriend farther downtown to her law firm job in the Seaport area, I thought I would grab some dinner, coffee, and check out the then-deserted, now-relocated Fulton Fish Market – deserted because all the fish-selling-flurry took place before dawn.

While roaming the hosed-off and brilliant cobblestone streets, I caught some movement out of the corner of my eye. Given the location, I naturally expected to see a grazing seagull, or an untroubled pigeon or rat finding breakfast. No.... It was a beautiful Blue Crab – there in the ancient gutter, its glistening shell backed up against the curb, claws up in defensive mode. But defending itself against...nothing. There were no gulls vulturing around, no street cats stalking it. The crab seemed to be committed to take on any and all comers. The City itself. Fate.

Illustration by Ted Enik

It must've slipped through the slats of a crate, or somehow escaped its wooden bucket of crab kin who'd been trucked-up from Delaware in the middle of the night.

Brave lil' guy ("gal" more likely – I've read that the larger females are the ones the restaurants want), she really touched me, poised there ready for the ultimate battle.

No doubt the adventure ended badly for Xena the Warrior Crab, but when I retell the story years later, after I rewind my tugged-on heart strings, I admire her instinct to not go down without a fight, to try to overcome the unbelievably bad odds, to survive.

Ted's story brought to mind the resilience of life in our "creature" population. Here, a small crab has the fortitude and bravery to take on the world – yes it's true that she had no choice of circumstances. But when we think of ourselves in such places, we often are the first to collapse or complain. Do we stand up to the world and show a defensive stance? Some do, but if truth be told, many of us do not have the "heart" that the animal kingdom members have – the survival instinct. Has it been bred out of us? Is this one of the lessons that our fair-friend creatures are here to teach us? Is this communication between human and animal/sea creature an important connection, or merely a lesson waiting to be learned or tossed?

Tip

We've come to learn: Always look at a circumstance and see it from a variety of levels and positions. And never toss a lesson, because you'll be doomed to repeat it.

Zorro

Angela, Dinah's daughter animal communicator, like most of us has had trouble communicating with her own animals. For some time, she'd been telling Dinah that she was seeing a black panther chasing the cats (in play) around her home. At first, she'd been reluctant to talk about such things (for the normal *what if I'm crazy?* reasons). She had been watching Dinah faithfully as she'd talked with animals (hers even) and then moved from that to talking with Light Beings. (Dinah can still hear her clearly saying, "Mom, you're sounding crazy again." But hey, what's new?)

At any rate, Angela called with a serious medical problem with one of her cats, Zorro. He was not only continually having bowel issues but had been diagnosed with a heart murmur. Now though constipation in a cat might not seem like a big deal, having it over and over with the cat becoming sicker and sicker put it right up there with any of the other horrible things that we don't want happening to our family members. The worse news was to hear the vet say that she did not expect him to get better and the pain would increase. Angela was forced to consider putting him to sleep to ease his horrible pain and loss of quality of life.

But, of course, with that comes the guilt – don't we all know that one well? Most anyone in the position of having to put an animal to sleep suffers with thoughts of not wanting to make the wrong decision or to hurt their loved pet. Angela asked Dinah to ask the Light Beings about the situation. She needed Zorro to know she loved him and she wondered what he thought about the situation – what did he want her to do?

At first, Dinah tried to communicate with Zorro directly, but she couldn't reach him, so she asked the Light Beings.

From the Light Beings:

You already know that Zorro is another of our sweet children. But we have no knowledge of his time. This is not of our doing. We will talk with him if you will wait. (A pause.)

He wonders who will take over his position. Who will watch over the others? No one else has the ability. Is there no medicine? he wants to know. He would speak with Mary please is his request.

Mary, though, was busy tending to another miracle. She was not available.

Light Beings:

Please then send her a message for a later time. These are the things she does and she knows how not to take one's illness in this case. Just tell her the issues.

Sometime later, Dinah was able to reach Mary on Skype. She typed in her best *Mission Impossible* voice: *Mary, your mission, should you decide to take it…* Then she explained the situation.

Mary:

This cat is a very old soul. He is like you, an instructor [meaning Dinah and her way of teaching others how to do something and then moving on to the next thing]. You know how I always talk about vitamin E? Well, I think she should give some to him. Cats are easy to give that pill to; it will help him a lot...at least for awhile. But he is very close…just not ready. He is searching for another to pass his knowledge along...cuz he is special. He has a chemical imbalance...not sure if it is due to kidneys or maybe his pancreas, which is stopping his digestive system from getting enough water to pass his stools easily – so the vitamin E should help.

Dinah:

So he doesn't want to move on now…

Mary:

It isn't that he doesn't.... He can't yet. Not just yet, but he's very close.

Dinah:

Okay, poor little guy. Does he not trust his information to one of the other cats in his family? Or is there someone else in the otherworld to help?

Mary:

They [the other cats in the house] are receiving it. He will do what he needs to do and then she will notice how quickly he will fail...she will know when it is time.

Dinah:

...And he knows that she loves him and such...

Mary:

He knows.... They always know. He is sorry that he is putting her through this. One more thing, ask her if she notices a stray coming around...that might be the one who will take the information he is worried about.

It wasn't long after, just a couple days, that Zorro became so ill that Angela felt the time being broadcasted to her very weakly. She of course went through all the things we all go through at this horrible time, and then eventually she moved on.

Sometime later, when Dinah had the opportunity to teach Angela more of the principles of animal communications, they were astounded to see Zorro jump up on her bed across from her and he said: "It's about time." [This was a strange class because as she was trying to interact with her animal guide at that time – a raccoon – an odd, what we now call, the *Woo Woo Snake* kept following her around. It was a cartoonish snake that said nothing other than "Wooooo, Woooo!" as it practically flew around her. It kinda freaked her out. She also had Bentley, Marla Brook's cockatiel (from *Psychic Pets*) sitting on her belly telling her that he would "poop" on her head if she didn't get things right. He was nothing if not persistent in his crotchety ways.]

Mary later said that she saw Zorro playing around with Angela from the Otherside, and now Angela trips over this spirit cat all the time, thinking it someone or something else. Angela says that this is indeed a case of playing around and that she hears him calling her as well – his famous baby cry.

So dear Zorro is doing well on the otherside and all is now as it should be with the remaining four cats at Angela's home.

Sometimes You Can't Even Tune Them Out!

Dinah learned early on from Anita Curtis, at her very first workshop about animal communications, that sometimes – no matter how hard you try – you just cannot tune out some animals. Bentley is the first culprit who jumps to mind, finding Dinah in the shower and determined to get his message out. That was in the days of shadows plaguing the bird's home in Hollywood (See Psychic Pets.) But it doesn't stop there and you should not be surprised to find that it might happen to you!

If there's one thing we've learned about those living in the Philadelphia region, compared to those living in other areas of the country, is that we are often loud and pretty persistent about our needs. And this is true not only of the humans in our fair area, but of the animals as well. We say that with conviction. Yes, the dogs, cats, birds, and fish have a hell of a lot to say in Pennsylvania as a whole, but those in the Philadelphia area are more likely than most to ram it down your throat if you choose not to listen and follow their directions to the letter. Hmmm…you may be thinking. Just how does that work? Very strangely.

Dinah tells us:

It all began on a day that was "one of those kinds of days." You know the ones. Nothing was going right – starting from early in the morning when the alarm clock had fallen in the dog's water bowl to the angry road-raged drivers on the highways to the harassing phone caller who was so certain that I was just the person who needed to buy whatever it was he was selling… and the list of the day went on and on. I was a fairly new animal communicator at the time, only having gained the skill a few months earlier and still learning the ropes as far as bringing animal voices into my head or keeping them out. I was not very good at either, but I was trying.

Of course, most people didn't believe that I could do such a thing at all and scoffed at me when I mentioned it. So I stayed quiet for the most part, talking about it only to those believers I could identify – which wasn't always easy. There were members, too, in my own family who were skeptics and this made things difficult at times for me. It was especially tricky to sit in front of a large fish tank at my son-in-law's home while a big fish stared out at me as it kept asking, over and over, to *send in the shrimp* (which was a kind of food that was being supplied). I'd made a casual suggestion that the fish looked hungry and maybe it would like some of the shrimp food and was told that it had already been fed. Conversation over. Except that the fish wasn't having *no* for an answer and kept asking for food over and over and over. *Send in the shrimp. Send in the shrimp.* Finally, after an eternity, my husband was ready to leave – my head was throbbing from the fish-talk and I could not wait to get out of there. But even as I was getting into the car, I could hear the fish calling to me. *Send in the shrimp!* I really needed to work on shutting out the animal talk.

From there, I had an errand to run, to pick up dog food at the pet store. The headache was still pounding and I continued to hear in a faraway whisper, *send in the shrimp* – didn't that fish ever give up? Into the pet store I went and back to the dog food section. I grabbed a bag and made for the counter. Next to the counter, in a cage that went from floor to near ceiling, was a parrot. And don't you know, he recognized that I could hear him right away. How, is beyond me, but he started screaming at the top of his lungs! And I don't mean squawking. This bird was screaming. And my head felt like it was going to explode. The lady at the counter started to ring me up. And the bird screamed even louder.

Finally, I yelled out, "What, what, what!?"

That startled the cashier and she jumped back a bit like I'd slapped her. Or maybe she thought I was going to rob her...

"Sorry," I said, stammering and pulling out my credit card. The bird was still screaming. "The bird…when did you change his water last?" My voice sounded like a demand. Her eyes went wide.

She answered in a whisper and a question, "Last night?"

"He wants fresh water," I told her and handed her my card.

She nodded. "I'll make sure to get him some…" She eyed me strangely and I could tell she was sizing me up as a nutcase (I'd seen the look a lot lately).

"I'm sorry," I said, "I have a headache. I'm an animal communicator and that's what he's yelling about. Wanting new water. And he wants it…. Now."

"Okay," she said.

I grabbed my dog food and left.

Spiderman at his author signing at the Pet Valu store.

Note: I did go back to apologize to the cashier at the pet store and she was okay with everything – she told me they "got all kinds in here." And they even invited me and my dog, Spiderman, back for an author signing once my first book came out! Ah, the life of an animal communicator in Philly.

Birds of a Feather Do Not Broil Together

Many of us don't think about the kind of story up next. The question is: How do animals that are slated to be our food react if they are not on the "chopping block?" Earlier, I discussed the theory that these animals are well aware of their circumstances and feel honor to be chosen for such a thing. It's hard to stomach that for most of us, and thousands – maybe millions and millions or more – are vegetarians for just this reason. (Of course, I remind you that there is communication with plants as well, so our nutrition values become still more difficult to fathom.) Still, in the next short story, a ray of love speaks to our hearts.

Meet Pam who thinks differently now about chicken.

I was fortunate enough this past Easter to acquire six chicks. They were supposed to be breeders, but I found out when they all turned white they were broilers instead. Well, a broiler is the chick you see with Frank Perdue [famous chicken-from-the-grocery-store king]. They live about three months and then become "nuggets!" Much to my dismay, I had broilers. They lived for about 6 months or more and were the best pets I have ever owned. They talked to me constantly, came running up to me when I went to their hen house and actually would lay their heads against my leg! They have all passed with the exception of the one rooster. He now resides at an Amish farm with red-colored hens. He is stark white and we can see him from the road when we pass by the farm. We named him T-REX and I smile every time we drive by.

Get a Job!

Lynn Zavaro, author of The Game of You, *had her dog, Lita, since she was just 6 weeks old! Her story is about animals as healers of the nature that plagues us. She says:*

Lita is a "healer" and has been trained by the Mãori healers of New Zealand to support people in body work. She also meditates with her every morning for forty-five minutes.

Lynn has a friend who is deaf and also a world-class animal communicator, Miranda Alcott. Miranda, being hearing impaired, uses her telepathic and intuitive skills to connect to animals and become more aware of their perspectives. She becomes a "bridge" to find resolutions to conflicts within human/animal families.

Lynn says: We did sessions with Miranda when Lita was having some behavioral problems and alas, what Lita told me, through Miranda, is that she needed a job! She was too bored and needed some work. Fixed things up right away!

Lita has been my healer in life, my little light. I used to live in Santa Fe, New Mexico, and one day having lunch at the restaurant El Farol, Lita showed up with her owner in a basket of puppies. She crawled up on my shoulder (3 weeks old) and I knew she was mine. (I had no desire to have a dog then.) Thus, her name became Farolita, Lita for short (the feminine for "Farol" – which means light and is also the word for the candles in brown bags you see in Santa Fe over Christmas). I believe she came to me to help me heal me and others.

Post Script

The wonderful healing dog, Lita, passed away in February of 2012, leaving all who knew her heartbroken and feeling the loss deeply. Lynn says:

With our pets there is only pure unconditional love. As a friend of mine once told me, "When we feel sad, we pet our pets..." In the act of giving love, we forget our own pain, our own loss, our own inner-tensions.... We learn from our pets how to love, opening our hearts to give for the sake of giving, asking nothing in return, like them. And in that act, we come to

know ourselves, the light in the center of our chest. Knowing our own beat, we succumb to that which life gives us. We are accepting, more detached, we allow the moment to unfold with an open hand....

In the loss of my beloved puppy (yes, she will always be my puppy to me...) I realized that the love of our pet is uncomplicated. Thus, love is abound, expansive, full and whole, even in death.... And in the saying of good-bye, love transcends...

An Invisible Barker

There are times when our animals like to communicate with us from the otherside, and people will see what they call ghosts or spirits and experience any variety of what they call paranormal phenomena. This is the same for our favored animal counterparts. Oftentimes, there is such a combined love between human and pet, that a passed animal friend likes to either stay behind to be with its owner or visit from time to time to check up on things. Such was the case for author Asandra's friend Ricardo.

Asandra was visiting her friend's house in South Miami for the first time. She says:

Ricardo, a photographer, was showing me his art collection and art books. On that day, I was wearing a white, cotton skirt that came down about two inches below my knees. While standing to look at a book that he was showing me, I felt something brush against the back of my skirt. Intuitively, I sensed a small, white dog. I turned around to look, but there wasn't anything there. This happened one more time while I was in the house, but I never said anything to Ric. Later, Ricardo and I went out to lunch. When we returned to the house, just as Ricardo opened the door, I sensed that there was a small dog barking and wagging its tail, excited and happy that we had returned. The feeling of this sensation was so strong, that I absolutely expected to see a dog on the other side of the door. At that point, I asked him if he had any pets, to which he replied, "No."

He did tell me, however, that he used to have a dog that he really loved named Jackie O. She was a small, white poodle and had passed away in 1977. Ricardo also told me that whenever friends visited who were "intuitives," they always sensed the white dog.

Photo Credit: Ricardo Coru

Photo Credit: Renee Harper

Cats and Dogs and Soul Mates and Loud Mouths

Author Renee Harper has had a houseful of pets for as long as she can remember. She tells three stories of communications that go beyond the physical to anchor in the heart.

The first story reminds us that the universe may have plans that we don't know about as to who comes into our lives and how that is to happen. There are no coincidences. If one is meant to connect with another, even under the strangest of circumstances, that incident will occur and both parties will be the better for it. Renee tells us:

I call Gizmo my soul mate. He belonged to a boss I had. She would always be going on vacations with her family and needed someone to take her dog. I always volunteered. We would go everywhere together when I would doggie-sit him. The dog park, pet store, around town. I hardly ever left him alone in my apartment. My boss pulled me into her office one day and asked me what I did with Gizmo when I doggie-sat him. I told her all the fun activities we did during the weekend. She in response told me that whenever I would return him he would mope for days and days. Depressed. It was no wonder. Her family kept him in the laundry room/closet. A space of only 3 x 3 feet. They hardly let him out to play or run around, even when her children were home. It was very upsetting and I started taking him home during the weekends even when my boss was not going away. She would bring him to work on Friday and I would bring him back to work on Monday. Then one day she pulled me into her office again and asked me if I wanted Gizmo, to keep. She told me that he got to big for her to carry around and she didn't want such a big dog. I told her yes and we never looked back! Because of his background Gizmo is a bit neurotic. He prefers small spaces and to be alone, though when at the dog park he will socialize a little bit with the other dogs. He is my best friend and I saved him as much as he saved me!

Renee now talks about another side of connection, that of gratitude and understanding. These are emotions felt not only by humans, but animals as well. It just takes a special person to understand. Renee continues with her stories:

Bamboo was found outside of my work office, in a drainage ditch. His mother abandoned him. I decided to keep him since he was so adorable. I went on a trip to San Francisco for a few days and a friend of mine took care of Bamboo. Upon my return he started acting very strange. He would not eat or drink or walk around. I took him to the vet and found that he had apparently swallowed a needle and thread while I was away. The needle punctured his intestines and he was going to die if they didn't do emergency surgery. The surgery cost over $2,000 and everyone who I have told the story to responded that they would have let the cat die. BambooBoo (as he is now called because of his boo-boo) knows that I saved his life. He is the most affectionate cat. He will sit with me while I am working on my computer and will wait patiently for me to pet him. He never leaves my side when we are together. He is a constant reminder of how animals can sometimes show more affection and appreciation then humans.

And lest we forget that our animals have something to say:

Our local animal rescue was having an art auction fundraiser. I had no intention to bring home another kitten that day, but went simply for the art. In the corner of the room where the fundraiser was taking place there was a huge cage with 6 kittens in it. All were adorable, except one was meowing so loud no one wanted even to look at her. She wouldn't stop meowing and it was ear-piercingly loud. By the end of the event, she was the only kitten left. I couldn't leave her there. So we adopted her and named her Elphaba after the character in *Wicked*. Elphaba took immediately to GupGup, our pug. Cuddling her, not leaving her side, sleeping with her. You name it, they did it together! To this day (4 years later) Elphaba can still be found cleaning GupGup's ears, walking around the house as a pair, and even taking a drink of water from the water bowl together. Elphaba has since stopped meowing as loudly, and it just proves that sometimes the loudmouth is the most affectionate, once you have heard what they have to say.

Lions and Cubs

Beth Wilder, a nature lover and fairy expert, was surprised to find out that wild animals recognize our desires and often respond – when they find someone worth responding to! Or maybe cubs are cubs no matter who is involved…and moms are moms. She relates:

When my twin nieces were about two years old, my sister and I liked to take them to the Louisville Zoo for outings. About that time, two cubs were born to a lioness, and everyone was flocking to the zoo to get a look at them.

There was a huge crowd outside the lion exhibit when we got there, and we waited quite a while to see the lioness and her cubs, but she was keeping them inside her cave where no one could see them. Finally, the crowd gave up and dispersed. Shortly after the last few people left, my sister and I started to leave the area as well. We suddenly heard the lioness growling at us as we pushed the stroller with my nieces out toward the exit, so we stopped and looked back. We were amazed – and charmed – to see her gently nudging her two cubs out of the cave to look at our two "cubs." I thought that was a really neat thing for the lioness to do.

Paw Paw, the Protective Familiar

One thing that we've heard over and over is that pet animals were sent as protectors for one or more humans in a family – on this side and the other. Whether it be an adult or a small child, protection was the word of the day.

KimmiJae, a host for Para-X Radio Network had a wonderful protection story to tell! She says:

Being a Witch, of course I had my familiar. As cliché as it was, it was my cat, Paw Paw, a domestic short hair. He chose me as an owner when my neighbor's cat had her litter. Paw was the runt; no one wanted him and I actually had another cat picked out. However, he was the most vocal and just had a certain "PULL" on me. I couldn't ignore him, so I thought: "Okay, there's a reason here. So, I'll take this noisy boy." He was my protector. He attacked anyone who raised their voice to me, lol, including my teenage daughter. He followed me everywhere, sat on my lap watching TV, watched as I did my rituals, sat close while I was meditating, and most importantly, looked right in my eyes when I spoke to him.

For twelve years I had him by my side, unconditionally. He has been my familiar in every sense of the word. But then he got diabetes and was failing fast. I held him and looked in his eyes…and they were different...dark and almost hollow.... It was time. I told him to visit often and for as long as he wanted...and I laid him to rest.

I cried for days, and about three days later, I was in my office and I heard the noise like I used to hear when he'd rub against the kitchen cabinets – that door being pushed against sound.... I KNEW it was Paw... He had a mouse that hung from a string in a doorway, which I took down the night he passed, and actually buried it with him under the maple tree in the back yard. I heard the toy hit the wall like it used to when he smacked it around. Then I started seeing "shadows" about his size running from room to room. The clincher was, when my daughter came home from college to visit over the holidays, she was walking from the kitchen to the living room, carrying a plate, and she stumbled as she tried to avoid tripping over something. She said, "Oh my God! Did you see that? What the hell!" I said, "That's Paw Paw! He hangs around." She said, "I swear this house is crazy! I come home and feel spirits in the house and now I'm seeing shadow cats!"

Paw still comes around now and then...not as often, but he knows he's welcome anytime and always.

A Short Lemur Acknowledgment

One of Dinah's two animal guides is a Lemur called Leo. She often talks about him to her daughter, Angela. One afternoon, her daughter was visiting the St. Louis Zoo and came to the exhibit that contained lemurs. One particular one came to her at the fence edge. Her daughter says: "He seemed as though he liked me and wanted to touch my hand."

It was a short moment, but when she told Dinah the story, she had a quick inkling that her own guide had prodded the small lemur to reach out in greeting. Our animal guides are not only aware of us, they are close to us.

Another acknowledgment, uncomfortable as it was to hear, came from Angela that said, "Most of the other animals seemed unhappy, like they didn't want to be there."

There is a fine line between being alive and feeling alive. Zoos have rescued many animals from certain death and the people there love them all dearly. Their lives revolve around helping, caring for, and protecting them. But sometimes, love is not enough. The trick is to understand when that line is crossed. But how does one do that if one cannot communicate? This is a reason why everyone should at least try to excel in animal communications.

Photo Credit: Angela Tallent

A Cat and Feathers

Dinah has found through years of ghost hunting and working with others in the paranormal field that we will oftentimes be given signs of things to come, or as signs of welcome, when finding feathers. This, then, is yet another way that animals (from the otherside?) communicate messages to us.

During one of Dinah's investigative interviews in beautiful Cape May, New Jersey, she was joined by a psychic medium to talk with the owner of the very haunted Elaine's Haunted Theater and B&B. There were many paranormal events taking place on that day (see Cape May Haunts*), but one in particular stood out.*

Dinah says:

As we were moving from one room to the next throughout the building, we came to one of the several areas of the business that held the spirit of the little girl, Elly. The psychic was getting impressions of her, our EMF (electromagnetic) meters were flashing red, and I was observing all of it. Suddenly, at baseboard level, I saw a gray…*woosh*…scoot around the corner of the door and out into the hall. "That was a cat," said the psychic medium. "Gray." I was so excited because, though I'd had many paranormal experiences, I'd not actually *seen* a ghost. Until that moment.

This then showed me that there may be certain connections between people on this side who love animals, animals on the Otherside, and very charged "haunted" areas. Did the cat want to communicate or change the world from his haven at Elaine's? Probably not. So, why then were we able to see it and why was it obviously cognizant of us (enough to run from the room)?

You may be asking: What do feathers have to do with this? A few days prior to our investigative interview, the psychic came down the stairs of her home to find a pile of purple feathers on the floor – all nice and tidy. She did not have a bird (much less a purple one). Immediately, there was the sense of welcome and knowledge that she would be visiting the haunted theater. To make matters more interesting, another part of our interview involved, yes, feathers. Ron Long, the owner of the establishment at the time, told us that in one of the rooms, two young women awoke to find their entire room covered in feathers.

There are times those on the Otherside just want to be heard and they ask us to listen.

Sad Messages

Sometimes the signs are for good welcoming reasons, other times we find that the messages may indicate something sad or foreboding to us – something we'd rather not know. Still, they are there to move us forward. Such was the case for Demonologist Katie Boyd when, quite suddenly, she lost her 22-year-old cat, Stevie. They'd been ready for the brother cat to pass because he was not aging well, but it had still been a surprise for them when Stevie took ill. Everything happened within a four-day period. Devastated, Katie says:

...two days before she became sick...I was in the bedroom and suddenly saw a bird flying through our kitchen. I caught the wing before it disappeared. Now the wing was bigger then a bat's wing and I saw it in a gray color. Right after I saw this bird, our smoke alarm would not stop going off, no matter what we did. It stopped when we brought Stevie's cremated remains back home.

Stevie was home; this was the communication that needed to be delivered to Katie that day. Whether the impact an animal has is on one person or millions, such effects allow our world to turn on its axis, to see the light at the end of any tunnel, to grow beyond selfishness to selflessness.

A View of the Otherside

Author Arlene Bice has found out firsthand that animals can see and communicate not only with those on this side, but also those on the otherside!

Cats are psychically sensitive animals. Maybe they are more so than other household pets. But then my girls, as I refer to them, are more than just pets.

Jane and Lizzie were named for the Bennet sisters in *Pride & Prejudice* and mirror their personalities, plus. Jane walks and reacts in a proud, quiet way. Lizzie is more feminine and more aggressive in a girlish way. She's not afraid to speak up and let me know what she thinks. All three of us knew that an unseen presence was living in our new home.

I first saw the image of an American-Indian Brave on a newly re-painted wall above the fireplace. I pooh-poohed the idea, chalking it up to my imagination. Jane, sitting at my feet while I sat in my comfortable wing chair reading in the library, saw him enter the room. I watched her fur go up on her back. She wasn't afraid of him but she was alerted.

One afternoon, both the girls lay on the floor in the library as I read. Lizzie was stretched out on the floor when she suddenly flipped over and the depression of her fur appeared as if a hand lay on her stomach. She was purring, enjoying it. He was playing with her. Jane sat by and watched.

Another afternoon, the two girls sat while I watched their eyes following him, turning their heads in synchronization as he entered the room through one doorway, walked across the room, leaving through another.

Aromas of roasting meat on an open fire wafted under my nose as I sat writing on my computer in my home office. All windows and doors were closed to keep out the cold winter air. I live in the middle of a forest with no neighboring houses in sight. Jane strode into the room, her food-sensitive nose in the air. She loves to eat, demanding to smell everything on my fork, rarely tasting it but always smelling it. Lizzie followed her in. They looked at me with question marks in their eyes as if to ask, "Where is it?"

He's gone now. I've not seen him for a several years. It seems that my Brave moved out when my ailing brother moved in with me for two years before he passed away. The girls and I miss his reassuring presence.

Photo Credit: Arlene Bice

Tom and the Horse

T. Stokes brings us another wonderful healing story that touches the heart.

The young woman held captive in a California garden shed for 18 years has been receiving therapy with horses to help her recover from the awful ordeal.

So learning to trust again is, for Jaycee, a very big issue, and horses are so important. Being among the best of animal healers, the red Indians are said to have put injured people with them for this very purpose, and young children with learning difficulties seem to respond particularly well to both horses and dolphins.

But beyond that, there are those who have no idea the value of animals in their lives – until they are immersed in a situation. T. says:

Many years ago, I would have to place men being released from long prison sentences in employment. One man had spent almost his whole life in some form of detention; his file was one appalling long catalogue of violence and theft, including debt collection for the notorious Kray twins. I anxiously thought I ought to meet this man called Tom before I placed him anywhere, and on asking him in for interview, he told me that he was only out a matter of weeks each time and would then be incarcerated yet again. But this time he was determined to stay out of trouble. Tom was a really hardened individual, but years of using the prison gym meant he was supremely fit and unofficial prison boxing champion. He had, though, shown a real interest in lovingly tending the prison garden.

Now this struck a cord in me because I had a call from an old lady the week before who had asked if I could get her help with her large overgrown garden. I very carefully asked this man if he would be willing to do this. So checking in with the old lady at the end of each day, she told me this man had worked tirelessly on the garden for his dinner and some tobacco, and slept happily in the large garden shed.

Now over the back to this old lady's large garden was an animal rescue field, and they had put a large horse there. The horse was the half-starved victim of severe cruelty and was there in the field to live out its last days.

No one could get near to the horse, and the guy from the animal sanctuary just could not get close to it for its medication. He'd chased after it each day, and had given up.

I had a phone call one day to get to the old lady's house quick. I am normally a good judge of character, but my stomach turned over as I thought about the old lady and the violent ex-criminal living in her garden shed.

When I arrived, she laughingly said, "Come and watch this!" The man from the animal sanctuary had come in early, and to his amazement, had found the horse, a large Suffolk Punch bred to pull heavy carts, trotting over to the fence and being patted and stroked by this very violent man.

Now if I had not seen this, I would not have believed these two were sharing such loving closeness, warmth, and care.

When the garden was finished and looking just beautiful, Tom the violent ex-criminal was asked to stay on at the sanctuary helping the dogs, and particularly, the horses that came in. Now this man may have been bad with people, but he sure had a way with horses. And recently, while in a dentist's waiting room, I flicked through a magazine, and who should I see as a top horse healer and therapist but Tom, the violent ex-offender.

The funny thing is: Who healed whom? And I could not help thinking it was the horse who healed Tom from his awful childhood and past life, not the other way round.

We should all take note of the power of love coming to us from the animal kingdom.

Bringing Murdock Home

Author Elinor Von Linden from Hobe Sound, Florida, had help communicating her loss of dear Murdock from the otherside. When animals are concerned, everyone gets into the picture! She tells us:

A few years ago, I had a beautiful silver-gray-haired cat named Murdock. Murdock was an indoor cat and never went outside. (As I used to say, "His little paws never touched grass.")

I went away overnight and left my husband to care for the cats. When I came home, Murdock was not to be found. I looked all over the house and no Murdock! As I had to work during the day, I spent every evening looking for him around the neighborhood. I knocked on all my neighbors' doors and gave everyone a flyer with Murdock's picture and my phone number. I would sit in the park across the street from my house and cry. I would talk to my father (who had passed) and ask him to send Murdock home. My dad loved cats as we always had a few while I was growing up. This went on for ten days.

On the tenth day, my daughter, Carrie, came home from college for Christmas break and brought her cat, Rockie. While we were sitting in the family room, Rockie kept going to the back door. I opened the door and there was Murdock! I know my dad had something to do with bringing Murdock home, and Rockie let me know he was there.

Photo Credit: Elinor Von Linden

Nike, the large black cat on the right, rocks with Prudence, the smaller cat on the left.

A Short Reiki Healing Story

Elinor Von Linden also has a story about Nike to share:

I had a beautiful black cat named Nike who was very timid and shy. She was afraid of everyone and spent a lot of her time hiding. (I don't think she was handled much as a kitten.) She was even fearful of me and my family. Because she was very fearful, she never purred.

After my first Reiki class, I decided to Reiki Nike. As I preformed Reiki on her, I heard this loud noise. At first I thought it was coming from the TV. But then I looked down at Nike and realized it was coming from her. That was the first time I heard her purr! As I continued to Reiki her, she continued to purr and she purred from then on.

Healing communications are often strong between animal and human when there is an already established love connection. For those with interest in developing that union, learning Reiki – even at just the first level – is an excellent way to begin animal communications.

Conclusion

We find ourselves at the conclusion of our stories, but not at the end of the time spent thinking of the impact that animals have on our living planet. Questions still arise regarding the why and the how of life's creation for the varied species. Somehow, we, as humans, have placed ourselves above those we call animals. After all, we think... (but maybe not all of us). We show our superiority, our gentle nods that reiterate that though we love the animals of our planet, we are much better than they are. Knowing that some of us believe this makes it difficult to end our journey on a high note. The stories here, for the most part, are loving gestures and offer wonderful knowledge that creatures other than humans have value and are deserving of love. But what of those of us in the world who do not "see the light"?

The lesson is driven home very clearly for me. Most recently, I had the wonderful opportunity to go on a cruise with my husband. While out in the ocean, which is so vast beyond words, I looked around and thought just how small we as humans (both in size and mentality) really are - invading "their" space and we can't even take care of our own. As I sat on the balcony outside our cabin, I was in awe as whales, dolphins, and an assortment of other fish swam by us - and they were adjusting to this large vessel. It occurred to me, that it was them making the adjustment to the intrusion and they were making it work. There was no striking, no violence, nothing other than trying to live in harmony with us. Why can't we learn from these wonderful sea creatures? They have been showing us all along how to make it work.

Questions like this pop up every day about animals, birds, fish, etc. - they always manage to survive. Their instinct, which they must follow to endure, is never questioned, they just act upon it. My Gracie, with her very first litter of kittens knew exactly what to do. One of the kittens came out lifeless. I was panicking when talking to the vet, wanting him to do something. But he said, "Let her be; she will know what to do." Gracie, who had always been a pampered cat, went into action and began licking the kitten's chest over and over. Suddenly, the kitten took a deep breath, and thanks to Gracie, I have my beautiful Daisy. How did she know? INSTINCT. She never questioned it, she didn't look to me, she just did it.

We can learn so much from just watching, observing, and living as they do. People say animals have no sense of caring or emotion or feeling - they are only dumb animals. I do not believe that to be true. A case in point involves two guard dogs in Japan. When the terrible tsunami hit, it was devastating and affected many. Along the beach, in a pile of debris, sat two dogs. One was hurt badly and couldn't move; the other sat next to the hurt one keeping watch over his friend. No one could get near them. They stayed with each other for days. The healthy one would not leave the other. He knew the injured one could not take care of himself and was willing to risk his own life to be with his partnering friend. Finally, both could be rescued and all was well. Point here is, no one told the healthy dog to sit by the other until help arrived. It was not only instinct, but also caring and feeling. They were buddies; he was not going to leave his friend behind.

So, getting back to me sitting on the balcony of this beautiful ship…I kept thinking…. We really need to take notice of these wonderful beings that we have been living beside all these years. They are a gift to us, they are part of us, and they will be there to teach us how to get along, how to live a non-cluttered life.

I cannot stress enough just how much animals impact our lives daily. But take note. Animals have been around a very long time. Information has been passed down to them and they have learned to listen. Maybe we should too.

Acknowledgments

We'd like to thank those listed below for their contributions to the stories, photos, discussions, and all kinds of varied support while we put this book together. We couldn't have done it without you!

Asandra

Author, Spiritual Advisor
www.ContactYourSpiritGuides.net or TheGuidedPath.net, Asandra.com

Asandra is an artist, writer, and full-trance channel with an international clientele. She channels Master Spirit Teachers exclusively with a focus on assisting individuals in the fulfillment of their souls' highest journey. Asandra is the author of Contact Your Spirit Guides. *She has also contributed to* More Hot Chocolate for the Mystical Soul, *by Ariell Ford,* South Beach Architectural Photographs, *and is the Spiritual Guidance Examiner for Examiner.com National. Her artwork is included in the book,* Miami Contemporary Artists, *and is in collections such as the Marriott Hotel chain.*

Arlene Bice

Author
http://arlenebice.blogspot.com/

Ms. Bice shares her home in the forest near Lake Gaston, North Carolina, with her two literary psychic cats, Jane and Lizzie (Bennet, as in Pride & Prejudice) while writing. She has four books on history and two books of true stories of hauntings in New Jersey published. In addition, her book of memoir and poetry is Life & Labyrinth. *Recently, her* Ghostly Hauntings of the Piedmont & Beyond *was published.*

Katie Boyd

Demonologist, Ghost Investigator
www.katieboyd.net

Katie is a Demonologist and occult sciences expert with a passion for the paranormal with both law enforcement and medical background. She has led over 400+ paranormal investigations, including poltergeist activity, UFOs/abductions, demonic hauntings, and more. Over her twenty years in the field Katie has seen, heard, and dealt with almost every form of entity and energy out there.

Marla Brooks

Author, Witch, Radio Host
www.marlabrooks.com/

In addition to the popular Ghosts of Hollywood *book series, Marla is a practicing Witch and hosts* Stirring the Caldron *on the Para-X Radio network. She explores all matters of witchiness, and talks to guests in both the pagan and paranormal worlds in order to help clear up misconceptions, discuss interesting topics, and above all, to have fun. She is author of titles,* Workplace Spells *and* Animal Spells and Magick, *and her first witchy fiction book,* A Bad Case of the Collywobbles *is also now available.*

Anita Curtis

Animal Communicator
www.anitacurtis.com/

Anita conducts lectures and workshops entitled "How to Hear the Animals," gives talks on the subject of animal communications, and visits facilities to walk through and communicate with animals for groups.

Joanne Emmonds

Author and Ghost Investigator
joanne@pcinj.org

Joanne was nine when she moved into a historic home already occupied by an active but friendly poltergeist. She has had a life-long fascination with the paranormal ever since. After establishing an engineering career and raising her special and gifted son, Joanne focused her spare time on her two life passions; ghosts and writing. Joanne now runs Paranormal Consulting & Investigations of New Jersey. She and her family share their current home with a much more subtle ghost as she helps other people with their hauntings and writes about life with the paranormal.

Ted Enik

Artist
www.tedenik.com

Author/illustrator Ted Enik has worked for most of the better-known publishing houses in New York as well as for the occasional advertising agency, greeting card house, and pharmaceutical company. He was honored to be tapped to (attempt to) fill Hilary Knight's venerable shoes by illustrating the latest Eloise hardcover, Eloise in Hollywood. *Currently, he is an artist for the popular* Fancy Nancy I Can Read™ *series. He is currently working on* WeeWitch, *a 4-seasons alphabet book of short poems which Ted co-wrote with astrologer and Tarotist, Beth Roth, and will be illustrating.*

Renee Harper

Author

Renee Harper has had a love for animals since early childhood. She resides in Bisbee, Arizona, with her menagerie of rescued pets.

Carolyn Giles

Tarot Reader and Collector

KimmiJae

Radio Host of *Reign Radio*

Jim O'Rear

Author, Actor

www.JimORear.com

Jim O'Rear has been involved in the entertainment industry for over 25 years as an actor, stuntman, and screenwriter, working on projects such as Day of the Dead, Star Trek 4, Lethal Weapon 3, Mortal Kombat, Nashville, *and more, opposite the likes of Martin Sheen, Burt Reynolds, Matthew Broderick, Robert Englund, Kevin Sorbo, and Chris Sarandon, just to name a few. Jim is also a successful screenwriter and novelist with several produced screenplays and published books. As a paranormal investigator, Jim has appeared in numerous documentaries, has written paranormal books (*Tennessee Ghosts *and* Hollywood Paranormal Films*), and is the founder of Celebrity Ghost Hunters.*

Anne McCurdy

Search Team Manager and Canine Training Director for Indiana Task Force 1

www.intf1curt.org/mission.shtml

Anne leads the Indiana Task Force One Canine Urban Response Team to provide highly trained, experienced, and certified search dogs and handlers to aid law enforcement and fire and emergency services in the search and location of missing individuals, live and deceased.

Lee Prosser

Author, Sensitive

Lee Prosser (deceased 2011) was a lifelong Vedantist (following Hindu philosophy), a paranormal sensitive who investigated the supernatural and UFOs, and wrote a popular column "Bide One's Time" about his experiences for www.ghostvillage.com. He will be missed, but he lives on through his works and love for animals.

Chuck Rineer

Photographer

www.crineerphotos.com

Chuck, after retiring from the U.S. Military, decided to pick up his camera and started photographing birds of prey, mostly Osprey and Bald Eagles. During this time, he decided to visit the Wolf Sanctuary of Pennsylvania. In the late 1970s, he was invited to the home of Bill Darlington (founder of the Sanctuary) where he had just been given two wolf pups. Twenty some years later, Chuck has assumed the responsibility of Staff Photographer for the haven for rescued wolves.

Mark Sarro

ITC/Paranormal Investigator, Musician, Radio Show Host
www.chestercountyprs.com; www.mkomega.com; www.unamasrecords.com; www.uberradionetwork.com

Mark Sarro is an independent ITC/Paranormal researcher, founder of the paranormal Chester County Paranormal Research Society, a musician with the band MK Omega, an A&R representative for Una Más Records, and a radio show host on the Uber Radio Network. He is the author of Ghosts of West Chester *and the co-author of* Haunted Gettysburg. *Mark has always been an animal lover and grew up living with dogs as a part of the family. Having his own pets he came to realize the special bond held between the pet and its owner. Zira (dog) and Smuckers (cat) were a big part of his life for 13 years. They both passed within 4 months of each other and have left Mark feeling the full impact of the loss of both of his beloved pets.*

Carol Starr

Author, Ghost Investigator

Carol has had an interest in the paranormal since she was very young. She saw her first apparition in an old farmhouse where her aunt and uncle lived. She began to have other experiences after that, along with building a curiosity of what happens after death. In 2005, Carol joined the ghost investigative research group, Chester County Paranormal Research Society, and has been investigating and researching the paranormal ever since. Carol is co-author of Haunted Gettysburg.

T. Stokes

Holistic Palmist, Lecturer in Paraspiritual Studies
http://t-stokes.co.uk/

T. Stokes is a spiritual counselor/healer, specializing in chirological diagnostic work from handprints, and guarantees satisfaction. He is often described as "Britain's psychic agony uncle."

Eva Sakmar-Sullivan

Visionary Artist and Environmentalist
www.stardolphin.com/

Eva spent time in both the Atlantic and Pacific Oceans swimming with dolphins and sea turtles. Her spiritual quest and travels to ancient and sacred places world-wide have been instrumental in transforming her work to its current form of expression. Eva's work can be used as a journey tool during meditation or deep relaxation. Her intent is that her work serve the highest good of the viewer.

Mark Schaedel

Student of Native American Studies

Elinor Van Linden

Author
Facebook: Sacred Spaces

Elinor "Cookie" Von Linden is an astrologer, Reiki Master, and a Feng Shui practitioner. She was born in Newark, New Jersey and moved to Florida in 1980. She is author of Sacred Spaces Meditation Cards.

Cece Wick

Reiki Master, Interfaith Minister, Author
Facebook: Cece Wick

Cece Wick has been dedicated to the spiritual path, walking the Ribbon of Light, for over two decades. As a Reiki Teaching Master, an EMF Energy Healer, an Ordained Interfaith Minister, and a Certified Hospice Volunteer, Cece brings love, comfort, joy and peace to those who are seekers, who have health challenges, life challenges, or whose life transition is imminent. As a minister, Cece lovingly brings Spirit to weddings, and has served as the Spiritual Leader for an annual women's retreat. Using her own mystical and magical experiences, Cece's stories aim to provide a transformative Divine connection for the reader.

Beth Wilder

Author/Artist

Beth Wilder is author of The Twilight Realm *and* Bugs In Shoes.

Ginny Zaboronek

Author, Teacher
ginny@inspiringbutterflies.com
www.inspiringbutterflies.com

Ginny is co-author, along with Marge Richards, of Inspiring Butterflies: A 27-Day Course of Self Discovery.

Lynn Zavaro

Writer, painter, teacher
www.lynnzavaro.com/

Lynn Zavaro has emerged as a leading voice in the art of guiding others to know themselves and create the lives they've always envisioned. Her game, The Game of You - An Interactive Way To Know Yourself, and Create The Life You Want *supports others in having fun while artfully discovering themselves on their creative life paths. She lives in Los Angeles with her loving husband, Shlomo.*

Our Personal Acknowledgments:

From Dinah

I would like to thank my family, not only for support in all my writing endeavors, but for their contributions to this "story." Carroll, Ariel, and Angela – you were more help than you know! Thanks to my co-author and explorer for this book, Mary: May this project be the first of your escapades in writing. And thanks to all who participated, making the animal impact on humanity real for many who have not been in the limelight.

From Mary

I want to thank my husband, Rick, for giving me support to move forward and realize I can make a difference. My children, Alecia and Ricky, for accepting all of this and living the love of animals with me. A special thank you to my good friend, Dinah Roseberry, for seeing the potential that I had and for encouraging me to have confidence. To my very good friends who believe in me: You are the best.

Thanks from both of us...

...to the friends and hard workers at Schiffer who helped make this project a reality.

A special thanks to Anne McCurdy both for her contribution to this project and her undying determination to connect with rescue dogs to further impact our world in the most wonderful of ways.

Selected Bibliography and Resources

"About Llamas." Pennsylvania Llama & Alpaca Association. http://www.plaa-net.org/about_llamas.htm. Accessed April 2011.

"Animal Hoarding: The Crazy Cat Lady, Explained." NBC News. http://today.msnbc.msn.com/id/23612118/. March 14, 2008.

AnimalSpirits.com. http://www.animalspirits.com/index1.html

"Bodach." Monstropedia. http://www.monstropedia.org/index.php?title=Bodach. Accessed April 2011.

Brainey Quotes. http://www.brainyquote.com/quotes/keywords/animal.html Accessed August 2011.

"Butterfly." Journey of the Heart. http://www.journeyofhearts.org/butterfly/bfly_myth.html (This site's resources: Gagliardi RA. Lepidopteral Symbolism: Lepidoptera Symbols Relating to Flight. Butterfly and Bugs Symbolism. Fourth Issue, November 1997. Available at: http://www.bugbios.com/ced4/butterfly_symbols.html. Cherry R. Native American Mythology. Butterfly and Bugs Symbolism. Fourth Issue, November 1997. Available at: http://www.bugbios.com/ced4/mythology.html. Mars J. Butterflies: Grace on the Wing. Kansas City, MO: Andrews McMeel Publishing, 1997).

"Butterfly." Wiki Answers. http://wiki.answers.com/Q/What_is_the_symbolic_meaning_of_a_butterfly. Accessed April 2011.

"Day of the Dead." http://www.carnaval.com/sf/DayoftheDead_links.htm. Accessed April 2011.

"Does talking to plants help them grow?" Pennyslvania State University. http://www.rps.psu.edu/probing/talkingtoplants.html. Accessed April 2011.

"Egypt announces find of ancient cat goddess temple." *The Seattle Times*, Seattle, WA 2010.

"From lion-goddess to cat-goddess." http://en.wikipedia.org/wiki/Bastet_%28mythology%29#cite_note-Velde164-2. Accessed April 2011.

"Ghost Cat: Island Hotel, Cedar Key, FL." http://hotels.about.com/u/sty/anightinahauntedhotel/ghoststories/Ghost-Cat.htm. Accessed April 2011.

Haunted Places. www.shadowlands.net. Accessed April 2011.

"How Many Pets Are in the U. S.?" Provided by PetPlace.com. http://webcenters.netscape.compuserve.com/homerealestate/feature.jsp?story=rainingcatsanddogs. Accessed April, 2011.

Huffington Post. "Dog Guards Injured Dog In Aftermath Of Japan's Disaster." March 3, 2011.

Highland Orchards, Inc., http://highlandorchards.net

Highland Orchards Farm & Market is located at 1000 Marshallton-Thorndale Road in West Chester, PA 19380; Phone: 610-269-3494

Indiana Task Force 1. www.intf1curt.org.

"The Plants Respond: An Interview with Cleve Backster" The Sun, Free Spirit. [Location unknown] July 1997. Translated into Czech, October 1998 for publication in Baraka [Location unknown] http://www.derrickjensen.org/backster.html. Accessed April 2011.

Roseberry, Dinah. *Ghost Hunters' Tool Kit*. Schiffer Publishing, LTD. Atglen, PA 2011

Roseberry, Dinah. *Ghosts of Valley Forge and Phoenixville*. Schiffer Publishing, LTD. Atglen, PA 2006.

Roseberry, Dinah. *Psychic Pets: Solving Paranormal Mysteries*. Schiffer Publishing, LTD. Atglen, PA 2011.

Schaedel, Mark. "Dolphin Plea." *Jen's Worldwide Network Newsletter*. August, 2010.

Stone, Robert B. Ph.D. *The Secret Life of Your Cells*. Whitford Press, Atglen, PA 1989

Stokes, T. "Animal Healer." *Psychic Magick*, email newsletter, Summer 2011 edition. P6

www.dictionary.com. "Empathy."

www.memorialhospital.org. "Grief."

Wells, David. "Your Wellsbeing." Issue 3, July 2010. Newsletter, UK.